SpringerBriefs on Pioneers in Science and Practice

Texts and Protocols

Volume 24

Series editor

Hans Günter Brauch, Mosbach, Germany

For further volumes:
http://www.springer.com/series/11446
http://www.afes-press-books.de/html/SpringerBriefs_PSP.htm
http://www.afes-press-books.de/html/SpringerBriefs_PSP_TP.htm

Konrad Raiser
Editor

Carl Friedrich von Weizsäcker: Major Texts on Religion

Springer

VDW

Knowledge and Responsibility
Carl Friedrich von Weizsäcker Society
Carl Friedrich von Weizsäcker Foundation

UDO KELLER STIFTUNG
FORUM HUMANUM

RUB

Editor
Konrad Raiser
Faculty of Protestant Theology (Emeritus)
Ruhr-University Bochum
Bochum
Germany

ISSN 2194-3125 ISSN 2194-3133 (electronic)
ISBN 978-3-319-03703-5 ISBN 978-3-319-03704-2 (eBook)
DOI 10.1007/978-3-319-03704-2
Springer Cham Heidelberg New York Dordrecht London

Library of Congress Control Number: 2014937683

The cover photograph of Carl Friedrich von Weizsäcker with the Dalai Lama was taken from the MSN Wissen website: "Meister des Wissens: Die größten Genies der Geschichte: Atomforscher des Friedens: Carl Friedrich von Weizsäcker", at: <http://db2.stb.s-msn.com/i/FD/69132C2964BCC60DB5C30DF4397A.jpg>. There is no reference to the photographer and no reference that this image is copyrighted.

Copyediting: PD Dr. Hans Günter Brauch, AFES-PRESS e.V., Mosbach, Germany

Printed on acid-free paper

Springer is part of Springer Science+Business Media (www.springer.com)

Books by the Same Author Published by Springer

- Ulrich Bartosch (Ed.): *Carl Friedrich von Weizsäcker: Pioneer of Physics, Philosophy, Religion, Politics and Peace Research.* Springer Briefs on Pioneers in Science and Practice No. 21. (Cham–Heidelberg–Dordrecht–London–New York: Springer-Verlag, 2015).
- Michael Drieschner (Ed.): *Carl Friedrich von Weizsäcker: Major Texts in Physics.* Springer Briefs on Pioneers in Science and Practice No. 22. Subseries Texts and Protocols No. 10 (Cham–Heidelberg–Dordrecht–London–New York: Springer-Verlag, 2015).
- Michael Drieschner (Ed.): *Carl Friedrich von Weizsäcker: Major Texts in Philosophy.* Springer Briefs on Pioneers in Science and Practice No. 23. Subseries Texts and Protocols No. 11 (Cham–Heidelberg–Dordrecht–London–New York: Springer-Verlag, 2015).
- Ulrich Bartosch (Ed.): *Carl Friedrich von Weizsäcker: Major Texts on Politics and Peace Research.* Springer Briefs on Pioneers in Science and Practice No. 25. Subseries Texts and Protocols No. 13 (Cham–Heidelberg–Dordrecht–London–New York: Springer-Verlag, 2015).
- See also the website on these five books with texts by Carl Friedrich von Weizsäcker, at: <http://www.afes-press-books.de/html/SpringerBriefs_PSP_C.F.v._Weizsaecker.htm>.

Acknowledgments

The editors express appreciation to the Carl Friedrich von Weizsäcker Society and to Dr. Bruno Redeker for permission to use translations from the volume *The Ambivalence of Progress* (Paragon House, New York 1988) and to reprint the three chapters from *The Relevance of Science* (New York: Harper and Row–London: Collins, 1964). Both publishers do not list these books any more in their programme nor do they list Carl Friedrich von Weizsäcker as an author. They were informed by Dr. Redeker as the Representative of the bequest of Carl Friedrich von Weizsäcker that the copyrights for all publications of Carl Friedrich von Weizsäcker were transferred to the Carl Friedrich von Weizsäcker Foundation. Neither publisher replied to the question whether they plan new editions of both books. Therefore the rights were retransferred to the Carl Friedrich von Weizsäcker Foundation, which granted permission to reprint all texts.

The financial support of the Udo Keller Foundation is also gratefully acknowledged. The editors further thank Ms. Ann Hentschel for providing translations for three of the chapters of this book. Special thanks are due to Prof. Michael Drieschner for his suggestions regarding the selections for this volume and for further assistance in producing the final manuscript. Finally, the major contribution of Dr. Hans Günter Brauch should be acknowledged: he has developed the idea for this whole series and has carefully accompanied and overseen the process of production of this volume.

Picture Credits

Most pictures in the first part of the volume were provided by the Weizsäcker family represented by Dr. Elisabeth Raiser who granted permission to use the photos. This includes the pictures on pp. 6, 9, 10, 11, 12, 15, 16, 17, 20, 22, 23, 24, 50, 112, 166, and 174.

All other pictures were provided by the Carl Friedrich von Weizsäcker Society represented by Dr. Bruno Redeker. Where possible, the original sources were identified. Thus credits are due to: to Deutsches Museum, Munich for the photo on p. 26; Archiv der Max Planck-Gesellschaft, Berlin Dahlem for the photo on p. 19; to Max-Planck-Gesellschaft/Filser for the Photo on p. 167; to HEKS/Jürg Gasser for the photos on p. 36 and 48; to the Vatican/Osservatore Romano for the photos on p. 122 and 126; and to the Templeton Prize — Clifford Shirley for the photo on p. 171. They all granted permission for the use of the pictures.

For the remaining pictures provided by The Carl Friedrich von Weizsäcker Society the efforts to identify the original sources and eventual copyrights remained unsuccessful. This includes the pictures on pp. 66, 74, 98. They are being used on the assumption that the rights as part of the Carl Friedrich von Weizsäcker bequest were transferred to the Carl Friedrich von Weizsäcker Foundation represented by Dr. Bruno Redeker who granted the permission for their use.

Contents

Chapter 1
Introduction

This volume is part of a small series that intends to make major texts by the German scientist and philosopher Carl Friedrich von Weizsäcker (1912–2007) available in English to a wider, non-specialized audience. Through his writings, many of which have already appeared in an earlier English translation, he has become known as a preeminent theoretical physicist and philosopher of science. Two of the volumes in this series present major texts from these two fields of his work as a scholar and academic teacher. What is less well known, at least for an English speaking audience, is his role as a leading public intellectual and political adviser, especially on issues of prevention of war, peace building, and a new international order. One of the volumes is dedicated to this area of his wide-ranging activities.

What is probably least known to a general international audience is his life-long concern for the religious dimension of human life. It has its roots in his early socialization and the historical context of his family. In a kind of summary of his insights in this field he proposed to consider religion under four aspects: religion as element of a culture; religion as foundation of a radical ethics; religion as inner experience; religion as theology. These four aspects also reflect different phases in the development of his personal concern for religion.

Since his early years he remained sensitive and receptive for the moral, existential, and metaphysical dimensions of religion. The moral and ethical challenge of religion found its clearest expression for him in the Sermon on the Mount, to which he returned again and again in his reflection. This concern became dominant in the context of his efforts to respond to the threats of a nuclear conflagration in a potential 3rd World War, which to him was a manifestation of the unresolved moral crisis of modernity.

The existential dimension entered his life initially through the encounter with the artist and religious mystic and prophet Hans-Henning von Voigt (1887–1969), better known under the pseudonym 'Alastair'. In his biographical accounts he attributes to Alastair the most important influence in the field of religion, including the intense interest in Asian religious traditions. Alastair, according to this account, was apparently infused with the conviction that a new time, the era of the

K. Raiser (ed.), *Carl Friedrich von Weizsäcker: Major Texts on Religion*,
SpringerBriefs on Pioneers in Science and Practice 24,
DOI: 10.1007/978-3-319-03704-2_1,

Holy Spirit, of the living presence of the divine, was about to begin; he was eager to gather followers into a kind of secret society to prepare the way for the advent of this new time. While C. F. von Weizsäcker distanced himself from Alastair after a personal crisis in 1952, he remained attracted by the praxis of an ordered religious life in community and developed the personal discipline of regular meditation.

Given his life-long interest in the philosophical implications of contemporary scientific developments he was bound to touch on the reality of religion as the core of the classical metaphysical tradition in philosophy. In this field the German philosopher Georg Picht was a life-long close counterpart. The concern surfaced at least indirectly in his first substantive publication on "The World View of Physics" (1st in German 1943; English translation London 1952), especially in the chapter discussing the "infinity of the world". It manifested itself with greater clarity in the famous lecture course on the "History of Nature", which he taught at the University of Göttingen in the summer term of 1946 (1st publication in German "Die Geschichte der Natur"; Göttingen 1948; English translation London 1951). As he explained in the opening lecture, the choice of the topic, which called for an interdisciplinary approach combining several disciplines of science with history and philosophy, was motivated by the conviction that the sciences had to develop the sense of responsibility for the social and political consequences of their explorations. This required moral and, according to him, religious insight. The last two of these lectures dealing with the condition of humanity, seen first from outside and then from inside, develop these convictions further and represent the first explicit and public statement on religion that we have from him.

In 1959, shortly after having begun to teach philosophy at the University of Hamburg, he was invited to present the "Gifford Lectures" at Glasgow University. The first of the two series of lectures in 1959 and 1960 was later published in English under the title "The Relevance of Science" (London and New York 1964). Dealing with "Creation and Cosmogony" it takes up and develops further the philosophical, historical and religious insights that were hinted at in the lectures on the "History of Nature" and in a way brings to full cycle the reflection on the human condition. This is his first explicit discussion of the aspect of religion as an element of culture, an issue that remained prominent in his later work. The second series on the "Philosophy of Modern Physics" (1960) presents a first survey of what became the focus of his specifically academic work on the philosophy of science during the years of his teaching responsibility at Hamburg (1957–1969); a more mature summary has been presented in the book "The Unity of Nature" (1st German edition Munich 1971; English translation New York 1980).

For the major part of his life, the most problematic aspect of religion remained its self-expression as theology in the form of doctrines and teachings formulated in conceptual language. He never questioned his Christian spiritual roots as a member of the Lutheran church and he frequently showed his familiarity with the biblical tradition. But he remained deeply skeptical about the institutional church with its doctrinal self-interpretation in the form of theology. During his years of teaching and research in Göttingen he sought to clarify his views in conversations with colleagues from the faculty of theology, especially the Old Testament scholars Gerhard von Rad und Walter Zimmerli, as well as the theologian Friedrich Gogarten. Through the mediation of a friend from earlier times, Günter Howe, he was drawn into an ongoing penetrating dialogue between physicists and theologians. And he frequently refers to the one face-to-face meeting and conversation that he had with Karl Barth, the leading theologian of his generation in European Protestantism.

His interest was of course not limited to the Christian tradition. Already as a student he had begun to study the main scared sources of the Asian religions, especially of Confucianism and Buddhism. He developed close personal relationships with representatives and teachers of Zen-Buddhism and of Hindu religious wisdom. He felt deeply attracted to the profound truth expressed in these traditions, which resonated with his own search more directly than anything in Christian theology. This is confirmed by the account he has given of the fundamental religious experience of being at one with himself and with the deepest ground of reality while visiting in 1969 the grave of the Hindu saint Sri Ramana Maharshi. After this existential encounter with the truth his skepticism with regard to religion as theology at first seemed to become even more pronounced.

A new appreciation of the task of theology developed for him in responding to repeated invitations to speak at Christian gatherings and to participate in advisory commissions for the churches. In particular, in 1976 he was invited to speak at a conference commemorating the 70th birthday of the German Protestant theologian Dietrich Bonhoeffer who was murdered by the Nazi regime for his participation in the German resistance movement. In studying the life story and the written legacy of Bonhoeffer and especially his search for a new language of theology to speak to a world come of age, C. F. von Weizsäcker felt the challenge to clarify his own critical understanding of the role of theology in the context of any religion.

This challenge became even stronger when, after 1985, he became involved in a process among Christian churches aiming at an authoritative Christian call for peace with justice, as Bonhoeffer had envisaged it already in 1934. He publicly presented the following appeal at a German Church congress in 1985: "We call on the churches of the world to convene a Council for peace. Peace is today the prerequisite for the survival of humanity. This peace is not secure. At an ecumenical Council, convened for the sake of peace, the Christian churches must collectively issue an injunction that humanity cannot ignore. Time is short." This appeal met with a very broad positive echo. In a small publication under the (German) title "Die Zeit drängt" (1986) he developed the social, political, ethical and theological reasons for this appeal. In a kind of personal diary which he kept during the years of these intensive discussions he noted down: "Now I'm obliged for the first time in my life to reflect seriously about theology". His latest books (Bewusstseinswandel, Munich 1988; Zeit und Wissen, Munich 1992) both include extensive chapters on religion which reflect the insights that as a non-theologian he had gained with regard to the role, the potential and the limitations of religion as theology.

The selection of texts for the contents of this volume has been made in the light of the foregoing interpretation of the different phases of his experiential encounters with, and his reflections about the religious dimension of human life. For obvious reasons, the selection opens with those passages from his 'Self-portrait' (Nr. 1), which refer explicitly to his encounter with religion in the course of his life. This self-portrait had already been published as the opening chapter in the volume "The Ambivalence of Progress" (Paragon: New York 1988).

Then follow three items from the first series of Gifford lectures at Glasgow, i.e. the lectures on "Creation in the Old Testament" (Nr. 2), "Christianity and History" (Nr. 3), and on 'Secularization' (Nr. 4). As indicated before, these lectures examine particularly the role of religion, in this case Christianity, as an element or a bearer of a culture. They have been prepared by the author himself in English and have been left untouched as to their stylistic and linguistic presentation, including the use of the Authorized King James Version for all biblical quotes.

The majority of the texts selected for this volume have been written in 1975/6; this was the period when he was challenged to deepen his reflection about religion, not least after the crucial personal experience during the visit to India a few years before. The key text from this period is the lecture on the theological development of Dietrich Bonhoeffer (Nr. 6). It was in this lecture that he presented for the first time his fourfold interpretation of religion, i.e. with reference to culture, ethics, experience, and theology. Analyzing the theological development of Bonhoeffer he discovered correspondences with his own intellectual struggles and came to a new appreciation of the role of theology. In preparation for this lecture he had once again studied the text of the Sermon on the Mount and reviewed the different types of responses and interpretations that this central text of the Christian tradition has received (Nr. 5). As a kind of follow-up to the Bonhoeffer lecture he went back to the struggle of Luther with the doctrinal positions of the Medieval Catholic Church and to the influence of Luther's emphasis on justification by faith on the formation of modern Protestant theology (Nr. 7). All of these texts have been published before in an earlier English translation in the volume "The Ambivalence of Progress".

Three more texts from the same period (and the same volume) have been included, i.e. the transcript of a long conversation on meditation (Nr. 8); a reflection on the moral problem of morality, occasioned by painful experiences in the encounter with positions of the radical left (Nr. 10); and a brief, but penetrating note about the relationship between physics and religion (Nr. 11).

Reference has been made before to the influence that the discussions around the project of a "Council of Peace" had on the further development of his interpretation of religion as theology. Since the understanding and practice of peace were the main objectives of these discussions, a selection of the contributions by C. F. von Weizsäcker have been included in the volume with his texts in the field of politics (s. Ulrich Bartosch (Ed.): *Carl Friedrich von Weizsäcker: Pioneer of Physics, Philosophy, Religion, Politics and Peace Research.* Springer Briefs on Pioneers in Science and Practice No. 21). However, he continued to evaluate the insights gained during these years. In Chap. 4 of his book 'Bewusstseinswandel' (Munich 1988) he presented his mature thinking on the religion. Under the title "The unfinished Religion" he responds to the questions: what is religion, what was religion, and to the concern about religion today and tomorrow. The last part of this chapter has been included here in an English translation made for this volume by Ms. Anne Hentschel (Nr. 12). She has also translated the brief piece on meditation (Nr. 9), which has been selected from Chap. 10 on "Experience of religion" of the last comprehensive account of his philosophy in the book "Zeit und Wissen" (Munich 1992). Otherwise this chapter presents a good summary of his insights and ideas developed over the previous decades, which are reflected in this volume in the texts mentioned before.

To conclude this selection his two major presentations on anxiety (Nr. 13) and on death (Nr. 14) have been included. They can serve as indications of his effort to apply the thinking about religion to basic features of human experience.

As mentioned before, most of the texts included here had already been published before in an earlier English translation in the book "The Ambivalence of Progress". In all cases the English translation has been thoroughly revised on the basis of the German original. The same is true for the translations by Ms. Anne Hentschel of the items Nr. 5, 9 and 12. For biblical quotations the New Revised Standard Version has been used with the exception of the items Nr. 2–4, where the author himself had followed the Authorized King James Version.

Berlin, March 2014 Konrad Raiser

Prof. Dr. Carl Friedrich von Weizsäcker and the editor of this volume, Prof. Dr. Konrad Raiser (1998). © The Weizsäcker Family represented by Dr. Elisabeth Raiser who granted permission to use this photo

Chapter 2
Self-Portrait

This self-portrait, instead of presenting the author's philosophic positions *in abstracto,* attempts to throw light on the path that has led him to these philosophic positions.[1] Do we learn and find our philosophy, or have we recognized it as that which somehow we have always known?

2.1 Preparation

For my twelfth birthday, in June 1924, I wished for and got a rotating celestial chart that could be adjusted to the day and hour. Shortly after that, we left Basel, where my father was German consul, for our summer vacation. We went to a secluded pension, Mont Crosin, in the Jura Mountains near Berne. The Swiss Independence Day was celebrated there, as usual, with fireworks on the evening of August 1. There was an outdoor ball for pension guests that began with a long polonaise. At one point the long line of dancers separated, and I managed to lose the young lady I was dancing with. I escaped from the crowd into the magical, warm, starry night with my celestial chart. The experience of such a night cannot be described in words; I can only give the residual thoughts after the memory has faded. God was present, somehow, in the indescribable magnificence of that starry night. Concurrently, I was aware that stars are balls of gas consisting of atoms and

[1] This text was first published in English in: C.F. von Weizsäcker: *The Ambivalence of Progress* (New York: Paragon House, 1988): 1–30; the German original is part of C.F. von Weizsäcker: *Der Garten des Menschlichen* (München: Hanser Verlag 1977): 553–597. All rights of the Weizsäcker Bequest were transferred to the Carl Friedrich von Weizsäcker Society, a nonprofit society whose executive chairman, Dr. Bruno Redeker, granted permission in March 2014 to use selected texts of Carl Friedrich von Weizsaecker. See also the website on this book on Carl Friedrich von Weizsäcker, at: <http://www.afes-press-books.de/html/SpringerBriefs_PSP_C.F. v._Weizsaecker.htm>.

K. Raiser (ed.), *Carl Friedrich von Weizsäcker: Major Texts on Religion*,
SpringerBriefs on Pioneers in Science and Practice 24,
DOI: 10.1007/978-3-319-03704-2_2, © The Author(s) 2014

obeying the laws of physics. The tension between these two truths must not be irresolvable. But how could they be reconciled? Is it possible to find the reflection of the glory of God also in the laws of physics?

I had begun, perhaps a year earlier, to read the New Testament. The truth of the Sermon on the Mount had deeply affected and disturbed me. If it was true, my life was false; perhaps all our lives were false. During a long talk with my mother I defended to the verge of tears the duty to refuse military service because of the commandment "Thou shalt not kill." During a night of intense religious experience, I had vowed to devote my life to the service of God. I cautiously added "If He should call me." I could only imagine becoming a minister in service to God, and yet I wanted to become an astronomer. My situation could perhaps best be described as "the moral law above me and the starry heaven within me." I still had to learn that, once we begin to listen, God is always calling; and later, that God is neither above nor within but that I am in God.

My family comes from Württemberg. My father became a naval officer in 1900, and switched to the diplomatic corps in 1920, two professions that serve the state beyond its borders. My mother, as a young girl, had rather imagined herself doing social work, not being a diplomat's wife; she fulfilled the requirements of his career *sans peur et sans reproche* with a humanity that is not always customary in that profession. Afterwards, during the 23 years of widowhood to the present, she cared for the old and sick in the family and stayed young by making excessive demands upon herself. There are ministers, scholars, civil servants, and military officers among my ancestors. The Weizsäckers were millers in Hohenlohe and, later, theologians. My grandfather Carl Weizsäcker, a jurist, was the last prime-minister of the Kingdom of Württemberg. On August 1, 1914, he said to his family, "This war will end with a revolution." He supposedly also said, "My fortune is lost." The fortune had been accumulated through decades of saving from modest civil-servant salaries to provide security for children and grandchildren. His prophecies proved accurate. After World War I, the corpulent, and rather short-tempered, old man took me, a little boy, for walks in the forests around Stuttgart. We gathered mushrooms for the family table, and he would occasionally stop to make some wise and witty remark on the occasion of finding "a parliament of mushrooms."

Carl Friedrich von Weizsäcker (1914) with his mother. © The Weizsäcker Family represented by Dr. Elisabeth Raiser who granted permission to use this photo

My father also liked to go on walks with me and talk about history and politics. In school, I was absorbed in history and indoctrinated my younger brothers and sisters. I thought about the best form of government and decided it would be a constitutional monarchy. In my fantasy land my friends and family were given roles, and there was foreign policy and wars; the leading statesman resembled, although contrary to my father's taste, Mussolini (of whom I knew hardly more than that he existed).

Carl Friedrich von Weizsäcker (1916). © The Weizsäcker Family represented by Dr. Elisabeth Raiser who granted permission to use this photo

As I grew older, two of my uncles played very important roles. My father's brother Viktor, the philosophical doctor, whose anthropological medicine is still not understood today, opened for me essentially important horizons. For decades, I worried about certain remarks by him, such as, "I believe the law of causality is a neurosis." My mother's brother, Fritz von Graevenitz, followed the family tradition, first as an officer, but after the war as a sculptor in accordance with his nature. He was a man who was newly enraptured each day by the magic and deep laws of beauty.

Carl Friedrich von Weizsäcker with his mother, his sister Elisabeth and his two brothers Heinrich and Richard (1926). © The Weizsäcker Family represented by Dr. Elisabeth Raiser who granted permission to use this photo

Then, my father was transferred to Copenhagen. At 14 years of age, while cycling in the Danish woods, I believed I could perceive the secret of life in nature around me, virtually down to spinning atoms. I began to think about the relationship between truth and beauty. For a while, the phrase, "beauty is subjective truth and truth objective beauty," seemed helpful to me. Afterwards, while having to make long, tiresome walks through the city during the damp, dark, and cold autumn, I thought that since truth was every-where, I should be able, with the right subjective attitude, to also find beauty everywhere. Thus, I discovered beauty in a spot of sunlight on a green copper roof and in the smell of fish in a basement grocery store. On an excursion to Lund on a magnificently bright, clear, sunny December day, I concluded that the natural laws that are valid for atoms are not those that are valid for visible bodies, but that the latter ones must be the consequence of the former.

Carl Friedrich von Weizsäcker with his father and his brother Richard (1926). © The Weizsäcker Family represented by Dr. Elisabeth Raiser who granted permission to use this photo

A few days later, my mother invited to our house the 25 year-old Werner Heisenberg. She had met him during a music evening where he played piano and amicably argued with her about the youth movement. This visit determined my direction in life. I was not only fascinated by the almost mad brilliance of the recent fundamental discoveries, which the rather shy, unassuming, blond young man radiated; his superiority in every field, including skiing and chess, was healthy and sobering for my enormous ambition and arrogance. He advised me and tested my mathematical ability, finding it barely adequate. I was soon convinced that theoretical physics was the science where I could find the answers to my questions about astronomy. When I discovered, in my final *gymnasium* year, that the field that actually attracted me most was called philosophy, I was tempted to study that. He was of the opinion that to practice philosophy relevant to the twentieth century one had to know physics; that one would learn physics only by practicing it and that one could succeed best in physics before the age of thirty and best in philosophy after 50 years of age. I followed his advice, studied theoretical physics, and have never regretted it.

A few months after Heisenberg's first visit to our house, my father was assigned to Berlin. I was waiting there for the beginning of the school year when, in April 1927, Heisenberg sent me a postcard saying that he was traveling through Berlin to Munich in a few days and that I could meet him at the Stettiner railway station in Berlin and accompany him in a taxi to the Anhalter station. In that taxi he told me of his yet-to-be-published uncertainty principle. Then he boarded the train and was gone. The following days were spent walking around Berlin reflecting about what he had told me. I had a continual dispute with my mother who upheld that freedom of will was a moral postulate to be defended against my physically defined determinism. I could never seriously consider the possibility of separating nature and the person. But now I saw that they could perhaps be united in a manner still mysterious to me. However, to achieve this, I first had to understand physics. I thought, "If I would now run into Einstein on the street, I would recognize him and, overcoming my shyness, address him and ask him what he thought about Heisenberg's theories." But he must have walked other streets, and I never met him.

Sometime later I suddenly realized with a brief, but profound shock that I was no longer tied to the religious faith of my childhood. I was less troubled by the well known conflict between church and science. To take the example of the debate about miracles: on the one hand one could apply the psychology of eye-witness accounts to reports of miracles; on the other hand, empirically verifiable natural laws are discovered under certain experimental conditions. In the presence of someone divinely inspired, still other laws could prevail. Finally, the tendency of confirming religious truth through breaches in the divine order of nature appeared rather impious to me. Historical criticism went much deeper. To have been raised a Lutheran cannot be considered an argument for the correctness of Lutheranism; to have been baptized a Christian is not proof of the falsity of Oriental religions; to have come from a religious tradition is not proof of the truth of religion. It still amazes me that there are educated, religious people who have not completed this simple thought process and are therefore helpless when confronted with the convenient arguments of antireligious rationalism. I was now in the difficult situation of having to regain and redefine the religious experience for myself.

This sketch of my development, of what I attribute to the circumstances of birth and upbringing, ends here. I will now pursue general problems, leaving aside much of my private life....[2]

2.2 Philosophy

.....When I was 16 years old, I was inwardly no longer connected to the Christian Church, but I came early to the conclusion that there is no sense in leaving the place where one has found oneself already; I have always been, and not unwillingly, a member of the Lutheran Church. But as much as the New Testament meant to me, the church meant little, or so it seemed to me to my disappointment. In the fields of my search, in ethics and mysticism, it did not challenge me; it made no demands on me, neither with regard to the Sermon on the Mount nor the Gospel according to St. John. At the university, I heard Johannes Wach's lecture on Asian religions. I read the Chinese classics in Wilhelm's translations, especially the short, gemlike texts of *Tshuang Tsis,* and the discourses of Buddha in K.E. Neumann's translation. One must read this so slowly that one breathes along with these teachings, "whose beginning soothes, whose middle soothes, and whose end soothes." Since then, I have felt more at home spiritually in Asia than in Europe, though fully aware of the deep cultural differences. I knew there are individuals there who see and who are.

The decisive religious influence by a living person came, for a long time, from Alastair, a highly gifted artist, a lover and mystic, a relentless self-examiner, and a man who always was in need of help. Today, 60 years after his brief fame, his drawings are again found on the art market. The way he played piano, sang, and danced, though, is lost to us. The countless translations, from which he tried for decades to make a living in pension rooms under flowers and silk, are scattered like his own verse. I never met the challenge that this relationship with a man two decades older than me meant to me, and I have painfully learned from it what I could hardly have learned otherwise. Like many who were close to him, I could never translate what was to be learned in his presence into thoughts or deeds, and thus I refer here to influences that were of less importance but easier to translate than his own.

[2] The original text of this self-portrait continues with sections dealing with "philosophical physics", 'politics' and 'philosophy' which have been left out in this selection that concentrates on religion. The text here continues with the final pages of the section on 'philosophy'. The full text of the self-portrait is included in the first volume of this series. (s. Ulrich Bartosch (Ed.): *Carl Friedrich von Weizsäcker: Pioneer of Physics, Philosophy, Religion, Politics and Peace Research.* Springer Briefs on Pioneers in Science and Practice No. 21).

Carl Friedrich von Weizsäcker with his future wife (1936). © The Weizsäcker Family represented by Dr. Elisabeth Raiser who granted permission to use this photo

In 1938 I was a guest at a retreat of the Lutheran St. Michael's Brotherhood in Marburg. I saw that I could not belong to this community, but I am grateful for the week of experiencing their liturgical order of the day. Liturgy and regularity impart something to the deep layers, not penetrated by willful reasoning, which, at least for me, is as necessary as food and drink. I have not been involved in any permanent liturgical community, because I have never found them to be connected to the modern consciousness that I needed. However, I have adopted the regular habit of daily morning meditation. I have never studied meditation formally, because I never met a teacher who challenged my intelligence enough, and perhaps because of my urge for independence. This is against the rules, dangerous and not recommended to others. I have never attempted to go to extremes in meditation, but rather have let come whatever came. Yet, I could not live without this regular retreat into quietness. The often held opinion that meditation is narcissism and exists in opposition to involvement with other people is an error whose existence is difficult to understand. To be sure, there are also, though rarely, contemplative ways of living that do more good to others through inaction or even without human contact than through any activities. The distortions of this rare gift and the many dangers related to the opening up of unconscious sources may have led to this error.

Carl Friedrich von Weizsäcker with colleagues (1937). © The Weizsäcker Family represented by Dr. Elisabeth Raiser who granted permission to use this photo

My relationship to the church normalized itself only after the war, when I learned how to appreciate everyday life. I am grateful for that decisive step to the American Christians, who were immediately prepared to help our country. The scientific colleagues abroad, the old friends (with the exception of Teller), initially investigated whether our relationship to the Nazis made us worthy of a new partnership; and who could blame them for this? But the Christians knew that "we are all sinners," and were there without questioning. Love creates love, exactly because it is undeserved. But I owe my relationship to theology almost exclusively to one man, who was not an academically trained theologian: the mathematician Günther Howe, whom I had met in Marburg in 1938. He talked to me about the relationship between the thoughts of Bohr and Barth and organized, for 10 years after the war, conversations between physicists and theologians. Here I learned the language of theology. Howe expected more from the church than I and thus suffered more from it; I learned from his suffering what it was all about. Now the scholarship of the Old Testament, the most believable aspect of learned theological work that I know, opened to me a 1000-year-long history that, especially in the calls to repentance of the prophets, still concerns us directly. It helped me to understand the historical situation of Christianity. I needed this when I sat with Howe in two church commissions, an ecumenical and a German one, about the ethics and politics of nuclear weapons. I learned to understand the judgments and attitudes of the church through the situation of its representatives. I learned to make reasonable political use of the good will of the church which maintained a distance from the general corruption and conflicts of interest. And it became easier for me to let quotations from the teachings of Jesus, always so vivid in my mind, slip into a conversation in such a way that they suited the concrete situation.

Portrait photo of the 27 year old Dr. Carl Friedrich von Weizsäcker (1939). © The Weizsäcker Family represented by Dr. Elisabeth Raiser who granted permission to use this photo

I could now also interpret the Sermon on the Mount to a certain degree, in the sense that I could connect it with my modern consciousness. It contains at least three layers of reality. The outermost is the universal ethic of the golden rule. This has certainly never been thought out more precisely than in the practical philosophy of Kant. It does not impose this or that commandment, but the form of the universality of commandments: "Let your behavior be guided by principles that you would want to be the guiding principles of all humankind." The Sermon on the Mount can be understood everywhere because it appeals to what makes people human. The *second* layer is the revelation of convictions as the place of ethical decisions. To fulfill the commandment does not mean that I do not actually murder my brother, but that I love him. This "but I say unto you" reveals our reality and its contrast even to the commandments consciously accepted by us. We humans have again and again avoided this unbearable tension. The specific danger of the church is the zeal of good works, also the good work of having the correct faith. Luther recognized the escapist aspect of this zeal. The works also cover deviations from literalism, from the strictness of the universal commandment. The scientific view of humans finally tends to side with his causally understandable psychic reality over against the commandment. But the commandment is the condition for the existence of human society; it is the truth, whose embodiment is peace. History is a chain of deserved suffering for disobeying the commandment. And the experience that the church calls penance could teach us, just as does the experience of psychoanalysis, that healing is not possible if we consider our psychological compulsions as our nature, if we do not distinguish ourselves from them and do not acknowledge the guilt as our own. Healing *is* possible; and with this we come to the *third* and intrinsic layer which in the composition of the Sermon on the Mount with good reason has been placed in the beginning as a prologue: the indicative of the Beatitudes. Without the imperative of behavior no society is possible, without the imperative of convictions no maturity of the person. But the world of the imperative is merciless; it drives the sensitive to desperation. The imperative is permitted only because there is the other reality: "Blessed are the peacemakers, for they shall be called the sons of God. Blessed are those who long for the spirit, for theirs is the kingdom of heaven."

This I learned to say and to appropriate to a certain extent. But until my fortieth year, there was "the moral law above me." I knew what was demanded of me but did not do it. I knew that humankind was moving towards catastrophe, and that only those who followed this way could help. With depression I realized that I became a "pillar of society." A personal crisis, where I became guilty toward people, liberated me. All of a sudden I saw the personal ambition in the self-imposed demand of perfection, in the postulate of helping the world. I experienced that there is an inner voice that teaches unmistakably and directly, if we only ask it with complete renunciation of our own will; it demands where it hurts the most, and it consoles where we would not have expected it. I limited my attention to the narrowest circle of duties and sacrificed my ambitions of knowledge and of politics. And then happened the first breakthrough to philosophical physics and political influence.

I put this report about the church and Christianity near the end of this essay because it contains experiences that provide answers to previous questions. The last question was whether Plato's philosophy of the Good can still be meaningful for us. The indicative of the Beatitudes can be interpreted to mean that being and duty come together and that the bad is the absence of being. The church tradition had accordingly incorporated Platonism and has presumably seen deeper than modern Protestant theology. But the consideration of Plato itself originated from the search for a doctrine of the subject in historical nature, which simultaneously understands that the nature we know is nature for subjects. One could call the empirical side of this science anthropology. However, an anthropology that does not know the experiences about which I have just tried to report, is certainly not a science. It can only be a registering of the more superficial contents of the consciousness of our era, inevitably connected with the wrong causal theories. Therefore, the anthropological relevance of art in modern times is so much greater than science, especially if the art is not 'committed,' that is, not inundated with well-intended contents of consciousness.

Prof. Dr. Carl Friedrich von Weizsäcker at the Institute of Physics at the University of Göttingen (1949). © Archiv der Max-Planck-Gesellschaft, Berlin Dahlem who granted permission to use this photo

This report about experiences must therefore be taken one step further. Twenty years ago, somebody visiting me in Göttingen said, I should seek contact with certain Indian sages, for the sake of the essential connection between Eastern wisdom and Western science. I spontaneously replied that I was not yet ready for that and I that the simple act of wanting to make contact would not be useful here. I was convinced that the Indians taught truth, and that, if their teachings were truth, the deeper self would make the move when the time was right. I held to this attitude for a long time. Martin Steinke, or Tao Chun, a German who became a Buddhist monk in China, personified to me the brilliant wisdom of Zen.

He became an older friend. I owe my meeting with Professor Mahadevan from Madras to Queen Friederike of Greece. He explained to me the Advaita teachings of the Vedanta. I immediately saw its closeness to, if not identity with, Plato's theory of the One. This world of things, of multiplicity, exists only for finite subjects; they, as parts of the world of multiplicity, exist only for each other and for themselves; in a truth that is being experienced only through meditative enlightenment, there is only one self. "This is you, O Svetaketu," as Hilmer addressed me in Copenhagen. At the beginning of 1968, the pandit Gopi Krishna from Cashmere visited me in Hamburg. In a split second I realized this was one to whom I could listen. I do not want to repeat what I wrote a few years later in the introduction to his book *Biological Basis of Religious Experience.* He was self-taught in yoga and certainly in Western thought, but in particular he was an eyewitness. His Kundalini experience concerns especially the physical realm, and it is painful that Western medicine has not yet taken notice of this. I assume that for today's science to be able to recognize what is experienced here it has to take the detour of physics.

Prof. Dr. Carl Friedrich von Weizsäcker at the Institute of Physics at the University of Göttingen (1952). © The Weizsäcker Family represented by Dr. Elisabeth Raiser who granted permission to use this photo

In 1969 I took a position in the German Development Service and used the opportunity for a fact–finding trip of several weeks to India. I could report much on the misery and the developmental work, and also about the fact that these poor have a much greater capacity for happiness than we rich. Actually, I was there for the sake of one particular experience. I met Gopi Krishna again, spent 24 h in the ashram of the highly venerated saint Anandamayi Ma in Vrindaban and one day at the *Aurobindo* ashram in Pondicherry. Mahadevan took me, with the kind support of the German General Consul Dr. Pfauter, to Kanchipuram to meet the head of the second largest Hindu community, the Shaivas (worshipers of Shiva). How might a high dignitary of the Indian church look? In a simple bamboo hut on a suburban street a little man with white hair sat on the floor and silently looked at us for a few minutes with powerful eyes; that was the audience with him. The utter cultural strangeness and the unquestioning human proximity were seldom as evident to me as in that moment.

Mahadevan's master had been Sri Ramana Maharshi. The latter, when he was 16 years old—the son of a Brahman, and a student at an American missionary school—had an experience of death. It became clear to him that "whatever is dying now is not I." A few months later, he fled to the old temple city of Tiruvannamalai, got rid of all possessions, lived only on what compassionate passersby and, later, venerators gave him, and remained in complete silence alone with the self whose presence he was. Eventually, he returned to an external routine, to regular eating and talking, and an ashram evolved around him at the foot of a holy mountain. He transmitted his inner blissful being silently, smilingly, and by asking questions. To those who questioned him because they saw God's presence in him he taught to ask: "Who is it then who asks 'Who am I?'" in order to lead them to recognize in themselves this same presence. He died in 1950, 20 years before my visit. I knew all this when I drove to Tiruvannamalai with Mahadevan.

Prof. Dr. Carl Friedrich von Weizsäcker with his daughter Elisabeth and his son Heinrich in Göttingen (1952).

The reader must excuse the fact that I do not actually describe what is indescribable, and yet speak about it; if I did not at least try, I could not have begun this autobiographical report. When I had taken off my shoes and stood before the grave of Maharshi, I knew with the immediacy of a flash: "Yes, this is it." Actually all questions were already answered. We sat in an amiable circle as a delicious meal was served on large green leafs. Afterwards I sat on the stone floor next to the grave. The knowledge was there, and within half an hour everything had happened. I still perceived my surroundings, my hard seat, the buzzing mosquitoes, the light upon the stones. But the layers had already been swiftly penetrated, the layers of which words can only give a hint: 'You'—'I'—'Yes.' Tears of bliss. Bliss without tears.

The experience brought me back to earth very gently. I now knew which love held the meaning of earthly love. I knew all dangers, all terrors, but they were not terrors in this experience. Should I now stay here forever? I saw myself like a metal ball that falls onto a polished metal surface and, after a moment's touch, bounces back to where it came from. But I now had become somebody completely different: the one I always had been. A younger German member of the ashram led me into a room where three older Indians sat. We greeted each other with a glance and sat together silently for an hour. My German friend made coffee for me in his room. Mahadevan came, we walked through the great temple section of the city. I slept in the very simple guest house at the ashram, and in the morning my friend accompanied me to a cave in the mountain, under large trees, where the Maharshi had lived for years, and where sometimes he settled the wars of the monkey kings in the trees above him. Then we traveled on. The experience left me with incredible gentleness during the next days and weeks. Its substance remained within me. Without it, I might not have survived the suffocating experiences of those years.

Prof. Dr. Carl Friedrich von Weizsäcker playing chess with his son Heinrich at his cottage in the Alps (1962). © The Weizsäcker Family represented by Dr. Elisabeth Raiser who granted permission to use this photo

My account comes to an end here. Obviously not, because the theoretical questions are answered, even though the reader will realize that their formulation has already aimed at the last experience just described. My account comes to an end because it is an interim report from a way station. Where does the bouncing ball go now? I may perhaps be allowed to say what I would still like to work on in the field of theory.

Prof. Dr. Carl Friedrich von Weizsäcker with his son Heinrich at his cottage in the Alps (1962).

I have promised a presentation of my thoughts on physics, and I still hope to do it with collaborators. Here, I still hope to get as far as elementary particle physics is concerned. If my approach is not wrong, this is the line where logic, epistemology, and ontology converge, and therefore the basis for an encounter with traditional religious metaphysics. The belief that one philosopher develops one philosophical system, and the other another, is representative of a past plateau of the history of philosophy, or perhaps rather an inclined plane. There is only one philosophy, and this is inseparable from concrete science, and many people work on its history. How far one's own contribution extends is difficult to foresee.

From a rather different side comes the pragmatic demand: to state, as far as possible, which course politics should take in the coming years in our country and around the world. The difficulty here is that a penetrating insight must be critical toward all existing practices and programs. It would be possible to focus only on expressing this criticism and exemplify it through one's own way of life, a difficult role if one takes it seriously. I have instead tried to propose practical policies, while aware of criticism, and I should give even more concrete and extensive advice in the future, from what my institute has taught me and continues to teach me. But in addition I still have a strong desire to develop at least the outline of a critical and therefore constructive theory.

This critical theory needs as its foundation the project of a historical anthropology. It approaches from the pragmatic side the same field that theoretical work is striving for. I will not be able to accomplish more than to indicate that field and to encourage those who want to work in it without shying away from scientific biology, leftist social theory, and religious experience. A few interpretations of Kant could be included here, and I would enjoy giving another lecture-course on Plato.

All this is the consequence of the theory, and should be done. We are only hiding if we do not think the thinkable. But we will all still pass through other doors. This theory, too, is probably only one of the great historical plateaus. One cannot think what one does not do. The external actions exist under the uncertainty of the political future. The inner work is to become receptive to new perception. Doing here means to let it happen.

Carl Friedrich von Weizsäcker in May 1951 at the General Assembly of the Deutsches Museum, Munich. *Source* Bildarchiv Deutsches Museum, Munich who granted permission to use this photo

Chapter 3
Creation in the Old Testament

The second lecture was devoted to cosmogonical myths.[1] I hope you still hear their wild and wonderful poetry ringing in your ears, and still keep in your minds the drama of the slaughter of gods by which in their view the world came about.

Now I beg you to listen to a text which you all know well and which I nevertheless want to read to you in its completeness:[2]

> In the beginning God created the heavens and the earth.
>
> And the earth was without form and void; and darkness was upon the face of the deep. And the Spirit of God moved upon the face of the waters. And God said, Let there be light: and there was light. And God saw the light, that it was good: and God divided the light from the darkness. And God called the light Day, and the darkness he called Night. And the evening and the morning were the first day.
>
> And God said, Let there be a firmament in the midst of the waters, and let it divide the waters from the waters. And God made the firmament, and divided the waters which were under the firmament from the waters which were above the firmament: and it was so. And God called the firmament Heaven. And the evening and the morning were the second day.
>
> And God said: Let the waters under the heaven be gathered together unto one place, and let the dry land appear: and it was so. And God called the dry land Earth; and the gathering together of the waters called he Seas: and God saw that it was good. And God said: Let the earth bring forth grass, the herb yielding seed, and the fruit tree yielding fruit after his kind, whose seed is in itself, upon the earth: and it was so. And the earth brought forth grass, and herb yielding seed after his kind, and the tree yielding fruit, whose seed was in itself, after his kind: and God saw that it was good. And the evening and the morning were the third day.

[1] This text was first published as Chap. 3 of the "Gifford Lectures" 1959–1961 in: C.F. von Weizsäcker, *The Relevance of Science* (New York: Harper and Row—London: Collins, 1964). All rights of the Weizsäcker Bequest were transferred to the Carl Friedrich von Weizsäcker Society, a nonprofit society whose executive chairman, Dr. Bruno Redeker, granted permission in March 2014 to use selected texts of Carl Friedrich von Weizsäcker.

[2] The German text follows the translation provided by the German Old Testament scholar Gerhard von Rad. For the English version the author has used the translation of the Authorized King James Version.

K. Raiser (ed.), *Carl Friedrich von Weizsäcker: Major Texts on Religion*,
SpringerBriefs on Pioneers in Science and Practice 24,
DOI: 10.1007/978-3-319-03704-2_3, © The Author(s) 2014

And God said: Let there be lights in the firmament of the heaven to divide the day from the night; and let them be for signs, and for seasons, and for days, and years: And let them be for lights in the firmament of the heaven to give light upon the earth: and it was so. And God made two great lights; the greater light to rule the day, and the lesser light to rule the night: he made the stars also. And God set them in the firmament of the heaven to give light upon the earth. And to rule over the day and over the night, and to divide the light from the darkness: and God saw that it was good. And the evening and the morning were the fourth day.

And God said: Let the waters bring forth abundantly the moving creature that hath life, and fowl that may fly above the earth in the open firmament of heaven. And God created great whales, and every living creature that moveth, which the waters brought forth abundantly, after their kind, and every winged fowl after his kind: and God saw that it was good. And God blessed them, saying: Be fruitful, and multiply, and fill the waters in the seas, and let fowl multiply in the earth. And the evening and the morning were the fifth day.

And God said: Let the earth bring forth the living creature after his kind, cattle, and creeping thing, and beast of the earth after his kind: and it was so. And God made the beast of the earth after his kind, and cattle after their kind, and everything that creepeth upon the earth after his kind: and God saw that it was good. And God said: Let us make man in our image, after our likeness: and let them have dominion over the fish of the sea, and over the fowl of the air, and over the cattle, and over all the earth, and over every creeping thing that creepeth upon the earth. So God created man in his own image, in the image of God created he him; male and female created he them. And God blessed them, and God said unto them: Be fruitful, and multiply, and replenish the earth, and subdue it: and have dominion over the fish of the sea, and over the fowl of the air, and over every living thing that moveth upon the earth. And God said: Behold, I have given you every herb bearing seed, which is upon the face of all the earth, and every tree, in which is the fruit of a tree yielding seed; to you it shall be for meat. And to every beast of the earth, and to every fowl of the air, and to everything that creepeth upon the earth, wherein there is life, I have given every green herb for meat: and it was so. And God saw everything that he had made, and, behold, it was very good. And the evening and the morning were the sixth day.

Thus the heavens and the earth were finished, and all the host of them. And on the seventh day God ended his work which he had made; and he rested on the seventh day from all his work which he had made. And God blessed the seventh day, and sanctified it: because that in it he had rested from all his work which God created and made. (*Gen. 1, 1–2, 3*).

How does this report compare with the cosmogonical myths?

Here, too, a *mythos*, a tale is told. We are still in the realm of mythology. Science has not yet broken the firmament of the heaven. Still a god acts as human beings act; still he speaks to man in human words.

Yet, how great is the difference! Not an infinite epic, but a limited, well-ordered report. No slaughter of the gods, no dragons of the sea, no building of the world out of the slain body of the enemy. A god who has no adversary builds the world like a house, he tends the earth like a garden. The language is not poetry but prose. It is precise and condensed. We should not be deceived by the sound of solemn antiquity with which we inevitably hear it. This language is in fact solemn, but it is solemn in the way of full and sincere lucidity. The age of reason is in the ascendancy.

The contents show the same features as the style.

The world is now seen as an understandable whole. The plan is laid out for completeness. The 6 days are precisely not naive mythology; they are a means of classification, precursors of Linnaeus' system. No doubt the writer believed in them literally; who would not believe in his own system? The writer really believed that God had created those categories which we use in describing living and dead things. How characteristic of the interest in classification is the repeated phrase "of his kind", "of their kind".

No doubt the elements of the mythical epic are used in several places. But there the myth suffers precisely what it describes as being the fate of the ancient gods: it is slain and its parts are used as materials in a new edifice. Some details of the Biblical report even seem to indicate a conscious polemic against the Babylonian myth. Thus the chaos of the beginning is still remembered: "the earth was without form and void"; *tehom*, the deep waters of the same verse, is the same word as Tiamat. But it is not Chaos which has borne the God; no: "In the beginning God created heaven and earth." God was there first and you cannot ask whence he came. That sun and moon are made on the fourth day, long after the creation of light, is most certainly not due to the naive ideas of one who would not have known that daylight comes from the sun. One should rather think that thereby light in its own essence is distinguished from the many sources of light of which sun and moon are but the greatest ones. Here the implicit polemic against the Babylonian astral religion seems evident. Old Testament scholars have told me that the omission of the names Sun and Moon for the great lights is not accidental. These names are the names of gods, but the two lights are no gods; they are lamps made to light the world and to count the days and the seasons.

In spite of these differences the narrative is a holy tale, perhaps even in a stricter sense than the myths. If it is a work of scholarship, it belongs to theology. Its centre is God, not the world. It explains the order of the world by telling how it was made by God.

Who can have written such a story? Modern scholarship has found out that this chapter of Genesis is to be ascribed to the so-called Priestly Code. This is the youngest of the three main sources of what later times called the Books of Moses. There are very good reasons for assuming that it was written down during the Babylonian captivity of the Jews, in the 6th century BC. Certainly centuries of slow formation can have preceded this final redaction. But the text as we have it seems to have been written in knowledge of, and in opposition to the Babylonian myth.

Probably the intellectual atmosphere of educated circles in Nebuchadnezzar's repristinated Babylonian empire was far more rational, far more 'modern' than the state of mind expressed in those time-honoured festival hymns which were then still celebrated by the priests. But the hypothetical influence of the rationality of an ageing culture would, I think, be insufficient to explain that solemn lucidity of Genesis 1. I am inclined to think that even the enlightened rationality of a nation grows under the shadow of its gods and passes away with them. Our text does not seem to mirror the restlessness of rational refinement but rather an unbroken peace with those deep powers in the soul which find no adequate expression in abstract concepts but which are reached by the image and the voice of the gods. The Jews have been able to interpret the world in a new, non-Babylonian way, because they believed in another god than the Babylonians. The God of the Jews has become the God of the Christians, and nobody will really understand modern Europe who knows nothing about the God of the Jews.

However these historical influences may have been, to the author of Genesis 1 the world seems understandable precisely because God has made it. Who is this God?

To the pious Jew for whom this tale is written his God is known from his childhood, just as all nations know their gods under whose protection they live. But Yahweh, the God of the Jews, is different from the gods of other nations. If I am to express it in the very words of the Jewish tradition, the Jews knew him as the God of Abraham, the God of Moses, the God of the Prophets. We now have to interpret these three names. Abraham stands for faith, Moses for the law, the Prophets for judgment. All of them stand for God's covenant with his people, they stand for God's promise. What does this mean?

Abraham stands for faith. He has been called by God: "Get thee out of thy country, and from thy kindred, and from thy father's house, unto a land that I will shew thee" (*Gen. 12:1*). That was not easily done in a time when the family and the clan were the only protection to a man. But Abraham listens and does what God tells him, he does it then and there. That is precisely what is called *fides*, faith. God promises him: "And I will make of thee a great nation… and in thee shall all families of the earth be blessed" (*Gen. 12: 2–3*). The promise is in the future. I must live my faith without seeing the fulfillment of the promise. In all that has been written by man, I know no more terrifying example of what faith demands than the offering of Isaac. But the promise is not broken and Isaac may live. "Abraham believed in the Lord; and he counted it to him for righteousness" (*Gen. 15: 6*).

Moses stands for the law. From the days of Abraham we see the Jews separated from the nations. Moses unites them to be a nation themselves by the law that God dictated him. God has created a nation by his covenant. From that time on this nation will not be able to live unless it keeps God's covenant. Thereby God is seen to be a God of separation. He severs his people, who have the law, from the other nations who do not have it. In his people he severs those who keep the covenant from those who violate it. Thus originates the opposition of good and evil, an opposition which cannot exist in the same strictness in polytheism where different deities rightly demand different acts. This opposition is not yet an abstract morality. To be good means to keep the covenant, and that means life. To be evil means to break the covenant, and that means death.

The prophets stand for judgment. To be good means life, to be evil means death. But the appearance of life is different. The prophets open the eyes of those who are willing to see the deeper truth. They do it, as everything relevant that is ever done, in and for a well-defined historical situation. Their teaching accompanies the political breakdown of the nation. They teach their nation to understand this event as God's chastisement. This is not to be understood as their cheap outward causality: God will do you good if you obey his laws. Even if the prophets had to use such language in order to be heard by the many, they really understand that to leave the law of God is the deadly tiling in itself, not its outward consequences. To leave God's law means to leave the source of life. Hence precisely the judgment makes clear the meaning of the promise: if you stay in faith you will live with God. I think this promise has been fulfilled. Like every state the Jewish state has finally broken down. Like no other nation the Jewish nation has survived the end of its political framework.

How is Genesis 1 connected with all this? The Old Testament is the history of God's covenant, written down for his people. How must a nation that understands itself in this manner understand the surrounding world? He who has the God of this covenant cannot have other gods before him. The first commandment is the condition under which Jewish life is alone possible. Thereby the existence of other gods is not at all denied. To a nation living in the time of Moses or of David such an idea would have been impossible. We see with our eyes every day how the gods of the heathen live and act in their nations. Just for this reason Yahweh is a jealous god. But as the Jews cannot be separated completely from other nations and particularly since the Exile mixed them with one of the greatest nations they cannot just disregard their gods. Our god is the god of good and evil, of life and death; our god is the only true god. This development is not only historically inevitable, it means a great step forward in religious cognition. The Jew cannot but realize that the other nations do not have the same understanding of life and death and thereby of good and evil which is given him by his covenant with God. The terrifying demand of a faith in one god has taken the Jew out of the relativities as well as of the monstrosities of polytheism; he has learnt now a moral lesson which he might not have been able to learn in any other way.

But if this is so, the Jews must understand everything in the world in the light of the true God. He is the god of a historical covenant. Therefore everything in the world must in fact belong to the history of this covenant. Thus the Jews have been the first people to understand the world as history. The book of Genesis weaves every memorable tradition of ancient times into this theology of history. Before the covenant with Abraham it places the covenants with Noah and with Adam, both of which include not just the Jews but all mankind. And before all these covenants it places finally that description of cosmogony which is compatible with the idea of the one true God. The history of creation tells how the stage was set for the history of God's covenant with man.

Hence its concise form, hence its contents. Yahweh may well have originally been a thundering wind-god of the desert; now even the role of the Lord of Heaven in a Pantheon, the role of a Marduk, Zeus, or Odin is too low for him. He cannot have grown out of the world, else there would have been other gods before him. No doubt historians of religion may place him at the side of the deities of light; yet he is not the light, but light was his first work. The history of creation which I read to you is followed, from Genesis 2, 4 on, by an older one, which scholars ascribe to the so-called Yahvist source; there God still forms man like an artisan out of the dust of the ground. In our younger text God creates in the same way in which alone he acts in Jewish life: by his word. "And God said: Let there be… And there was…"

Since God now is so highly exalted above the whole world, everything in the world is of the same nature: it is a creature of God, it is not God. Thus God himself has deprived the world of its divinity. De-mythologizing has become a common word in our century. If mythology means that our thought is under the domination of the gods, it is precisely the faith in God which has de-mythologized our thinking from the time when the Old Testament was written. That is why I called this history of creation an anti-mythical myth. We will have to watch this process through Christianity till our time.

Everything in the world is God's creature. But one creature is distinguished above all others: man. He is made in the image of God.

I should not doubt, looking at the question historically, that man-shaped idols have made this idea possible. But then, how did man-shaped idols come about and what did they express? A fetish is divine, being the thing it is; it does not need a human face. Animal powers of the soul may incorporate themselves in animal-shaped deities. But man, learning to understand himself as a responsible person, has seen a man-like person in the god who made him such. The God of the Old Testament speaks to me. He says 'thou' to me and hence I can say 'thou' to him. The personal God is the God who made man to be a person. "I have called thee by thy name, thou art mine" (*Isaiah 43: 1*). The Jewish religion has finally reduced all our relations with God to this speaking and hearing. Even his likeness could no longer be made by hands. The man-like image of God was hidden or forbidden; he who sees God must die.

Thus we, mankind, who can say 'I' and 'you' and 'we', are made in the image of God just in this personal quality. Hence it is meaningful that God should tell man to subdue the earth. Expressing the same thought less mythologically we may say: if we have faith in God we are no longer the slaves of the gods. The gods are the powers of the world, within and without ourselves. If we have faith in God we are free in the world. This is why Jews and Christians have martyrs, witnesses of this freedom. The same freedom from the gods enables man to a re-shaping dominion of nature. In this sense I think even modern secularization, even scientism cannot be understood without the background of which I am speaking now. I shall come back to this point in later lectures.

But now we approach a dangerous possible misunderstanding. We have now so clearly distinguished the God of the Jews and the Christians from all pagan gods that we may have come close to the concept called the God of the philosophers by Pascal. The God of the philosophers is a pure spirit, the highest entity, the first cause of the world, all-knowing, almighty, all-bountiful. Wherever in later times we read about God in Jewish, Christian or Mohammedan philosophy and even theology, we find a concept of God that can somehow be described by the attributes just mentioned. It is very important to see that this is not the God of the Old Testament.

Let us begin by one particular attribute, God's spirituality. It is true that the Old Testament speaks of the Spirit of God. But then it does not mean an immaterial principle as opposed to matter. The very concept of matter is foreign in the Bible. Spirit in the Old Testament is breath. Some modern interpreters who know that have even been tempted to translate the *ruach Elohim* of the second verse of our text not by "the Spirit of God" but by "a storm of God". But the sounding voice, the word itself is breath, too. Every animal lives only as long as it breathes. Thus my breath is my life, it is myself as a person, appearing to the senses. Hence the Spirit of God is the Divine Life by which men or women at times can speak God's word and do God's deeds.

Later theologians who were usually spiritualists have read their spiritualism into the Old Testament, thinking that the Bible, being the truth, could only contain what seemed to be truth to them. Even if we take a more critical view we may feel tempted to think that by eliminating all that is mythical from the Old Testament's understanding of God we would arrive at the Most Perfect Being of metaphysical theology. This is not so. In pursuing such an argument we do not arrive at a consistent concept of God but at a paradox.

We can illustrate this by returning once more to the history of the creation in Genesis 1. It is one of the most rationally consistent parts of the Bible. I have tried to explain the line of argument that seems to have led to its rather late addition to the body of Biblical tales. Just this consistency has made it a possible starting point of the metaphysical theology of which we will speak in later lectures. But the very same consistency makes it no more than an outpost position of the Bible. I said it was the setting of the stage for the covenant. But the setting is not complete before the serpent has appeared. In fact the tale of the fall is older and theologically more important than the history of creation. For where would we find ourselves if we had only the history of creation? It seems to teach the making of all existing things by an almighty, good God. Later theologians were able to interpret it as teaching the creation of the world out of nothing *(creatio ex nihilo)*. All that God made was very good. What else could thence be expected but a history which would be the sheer fulfilment of God's will? But not even philosophers have dared to interpret the real world in such terms; and most definitely this is not the view of the Bible.

To the Bible God is the God who divides good from evil. The good is his work. The evil is not his work; it is rejected by him. Yet the evil is real. Man goes on doing the evil. This is the true picture of the world. Therefore the myth of creation had to be followed by the myth of man's fall. The great poet who wrote this myth in Genesis 3 showed a deep understanding by not saying whence the serpent's subtlety originated. He said precisely as much as he could say and kept silent about the unknowable. In any case only with Adam's and Eve's fall could the real history of the world begin, that history which is a struggle between God and man from the outset. Man has deserted God and deserts him daily. The dialogue between God and man which is the moving force of history has no other theme but to redeem man from his desertion. The covenant is an action in this battle, the desertion from God is death, and God calls me to life whenever he speaks to me. Faith means nothing but the trust in this call.

In the last sentences I have tried to express the true meaning of the Bible on this point. Perhaps this meaning has up to this day never been expressed without the use of some myth; we do not know whether it will ever be possible to express it otherwise. But one thing is certain: whoever for the sake of consistency or of harmony removes this battle from the centre of the picture, will exclude precisely that experience of God from his thought, on which Judaism and Christianity rest.

Prof. Dr. Carl Friedrich von Weizsäcker at the European Ecumenical Assembly in Basel (1989). *Source* HEKS/Jürg Gasser; permission to use this photo was granted

Chapter 4
Christianity and History

Christianity is, historically speaking, the religion founded by Jesus who was called the Christ.[1]

I dare not speak about the man Jesus. I do not feel equal to that task. But *I* owe it to you to make my personal position clear. I am a Christian, or I should rather say: I try to be a Christian. This is not a traditionalist's position. I have even found much of the Christian tradition, both in thought and in life, difficult to understand, and some of it impossible to follow. But, if this is a possible English phrase, *I* have been hit by the word of Christ. In a way this word has made life impossible to me; the life I might have lived without it has been destroyed by it. In a way it has made life possible to me; I am not certain whether I would have found a possible way of life without it at all. His word means his teaching. But since this teaching refers to life, his word should be understood to include the reports we have about his life, his death, and that mysterious event of which his disciples spoke as his resurrection.

But these lectures are not intended to contain personal confession but critical thought. I had to make my own position clear for what you may call a methodical reason. Let me use science as an example. Science rests on experience. If you discuss a scientific theory it is not necessary that you should have made the relevant experiments yourself. But you must know what scientific experience is, and you ought to make it quite clear to what particular experiments you are referring. Religion equally rests on experience. Experience, in religion as well as in science, is always particular experience. It will be difficult to understand any religion if you do not have the experience of living within a particular religion. In religions like Judaism and Christianity the relevant experience cannot be had

[1] This text was first published as Chap. 4 of the "Gifford Lectures" 1959–1961 in: C.F. von Weizsäcker, *The Relevance of Science* (New York: Harper and Row–London: Collins, 1964). All rights of the Weizsäcker Bequest were transferred to the Carl Friedrich von Weizsäcker Society, a nonprofit society whose executive chairman, Dr. Bruno Redeker, granted permission in March 2014 to use selected texts of Carl Friedrich von Weizsäcker.

K. Raiser (ed.), *Carl Friedrich von Weizsäcker: Major Texts on Religion*,
SpringerBriefs on Pioneers in Science and Practice 24,
DOI: 10.1007/978-3-319-03704-2_4, © The Author(s) 2014

without a personal decision. The act of the decision, if it really happens, is itself an integrating part of the experience. Hence, if you want to speak clearly about religion you will have to indicate what decision you have taken.

A decision chooses one way and rejects the other ways. Hence in opening an experience it cuts me off from other experiences. Making a decision cuts me off from some fields of religious truth, for the other religions contain their truth, too; avoiding a decision, however, will cut me off from the whole field of that truth which cannot be had by just looking at it. I do not say that this necessarily means a final cut-off. There is, I believe, an internal structure in the realm of truth, and, having gone through the experience of decision, we may understand in the end even that truth whose experience we rejected. But this process may well last a few 1000 years.

The Christian Church originated as the community of those who, as I have tried to express it, were hit by the word of Christ and who tried to live a life according to what they understood to be his will. The language in which they expressed the experience in so living had to be a language they knew. This was to a large extent a language which we would call mythical today. Here is the problem in our interpretation of Christianity. We want to express our experience in a language that comes naturally to us. Are we to abandon myth for the sake of what we take to be rational sincerity and to run the risk of thereby losing the very experience that was expressed in the mythical language? Or are we to retain the mythical language? Can we use it, knowing it is mythical? Are we so sure that we know what is myth and what is not myth? Do we even know sufficiently well what we mean by myth? I have not so far given any rational definition of myth in these lectures.

We shall come back to these questions at the end of this series of lectures. Now I shall try to describe Christianity as it has understood itself, and for this purpose I shall use the concepts introduced in the earlier lectures.

Christianity originated within the Jewish religion. This religion I have tried to describe by the three concepts of faith, of law, and of judgment. These three concepts are retained in Christianity but with a new interpretation. I think it is not wrong to say that faith, law and judgment have been transformed into the Pauline triad of faith, charity, and hope.

Faith is still faith in the God of the Jews whom Jesus taught his disciples to call our father. But it is now at the same time faith in a man, in Jesus himself. This was not less paradoxical then, than it is today. It is the paradox which is expressed in the Gospel of St. John by saying: The word was made flesh (1.14). The word (logos) in the language here used is the power of God by which the world has been created. Flesh means man, living in this world. If I wanted to interpret the verse precisely I should have to speak of the Gnostic meaning of the two concepts word and flesh; there they are opposites exaggerating Plato's distinction of reason and matter into a strict dualism, and the paradox of the verse becomes even more stringent then. But even without such an interpretation, staying within the Jewish religion, we can see that the verse identifies God's creative power with a creature that has lived once in history. Thus it draws even creation itself into that battle between God and man which is history. The appearance of one man in history who did not fight against God but who really fulfilled his will is understood as a second creation. The meaning of this second creation is seen in the transformation of law into charity, and of judgment into hope.

The Jewish law meant charity. When Jesus says: "Thou shalt love the Lord thy God with all thy heart, and with all thy soul, and with all thy mind... and thou shalt love thy neighbour as thyself" (*Matt. 22. 37–9*), he is quoting from the Old Testament (*Deut. 6. 5 and 19. 19*). He rightly says: "On these two commandments hang all the law and the prophets" (*Matt. 22. 40*). Still, the fourth gospel quotes him as saying: "A new commandment I give unto you, That ye love one another; as I have loved you, that ye also love one another" (*John 13. 34*). To explain this apparent contradiction I do not want to revert to the differences between Matthew and John, indisputable though they are. I think the words "as I have loved you" say what is new. Christians have at all times understood Jesus to have been the man who by his life has shown the actual possibility of perfect love; and this they mean by saying that the word of God has been made flesh. The Sermon on the Mount does not speak in the grammatical form of the imperative as does Moses' law; it does not say "Thou shalt not..."; it speaks in the indicative: "Blessed are those who..." This is why the life of Christ is an integrating part of his word. I can evade a law, saying that I am not able to fulfill it. Can I evade the example of a life?

Judgment and hope have been close together even in the Jewish prophets. Historically the early prophets, who spoke when the Jewish kingdom still existed, spoke of the impending judgment; they had to destroy a false hope. The prophets at the time of the exile spoke of the impending hope open to those who accepted the present judgment. This hope the Jews interpreted by expecting the Messiah who would restore the kingdom of David. The Christians said that this hope was now fulfilled, that Jesus was the Messiah; this is what the name Christ means. This is the same paradox as before. It is true, it crushes the hopes of the Jews, for the visible kingdom of David has not been restored. Jesus himself has reinterpreted the idea of the coming kingdom: "The kingdom of God cometh not with observation: Neither shall they say, Lo here! or, lo there! for, behold, the kingdom of God is within you" (*Luke 17. 20–21*). The words "within you" are probably a mistranslation, but a profound one, the original meaning being: "it is among you". In any case: it is here, you have no longer to wait for it.

But if this were all that was to be said about the expectation, I should have spoken of fulfillment, not of hope. We have to learn that the words hope, judgment and fulfillment are ambiguous in the New Testament. John speaks clearly of the judgment as fulfilled in Christ: "He that believeth on him is not condemned; but he that believeth not is condemned already... And this is the condemnation, that light is come into the world, and men loved darkness rather than light..." (*John 3. 18–19*). Equally clearly Matthew speaks of the coming judgment and the second coming of Christ.

This ambiguity seems to be inevitable. Christ has opened a way for every human person. To go or to miss it is that division between light and darkness, life and death, which is judgment. But if human persons are changed, history cannot stay unchanged. The fulfillment offered to the single person is the hope opened to mankind, the hope of history. This hope again means separation. "Think not that I am come to send peace on earth: I came not to send peace, but a sword. For I am come to set a man at variance against his father, and the daughter against her mother, and the daughter-in-law against her mother-in-law" (*Matt. 10. 34–35*). This is no longer the separation of one nation from all other nations. The separating line now goes through all nations and through all families. While the separation of the Jews from the other nations—whatever may have been the historical facts—was understood to have been effected at once, by the covenant of Sinai, this final separation is a process. In the parable of the tares among the wheat the servants ask whether they should gather up the tares but are forbidden by the householder, "lest while ye gather up the tares, ye root up also the wheat with them. Let both grow together until the harvest: and in the time of harvest I will say to the reapers, Gather ye together first the tares, and bind them in bundles to burn them: but gather the wheat into my barn" (*Matt. 13. 29–30*). And Jesus, explaining the parable himself, says, according to the gospel: "the harvest is the end of the world". History will come to an end.

With this expectation Christianity has entered history. Again and again Christians thought the last judgment to be close at hand. It never arrived, but history was steadily and incessantly transformed by those who waited for nothing but for its end.

As the first step of this transformation we may consider how the shattered hopes of the disciples, who had expected that Jesus would bring the judgment in his lifetime, were revived in a changed way by the report of his resurrection and by the profusion of the Holy Ghost seven weeks later. Whatever may have happened then, what they had wrongly expected to come immediately and from outside, they now knew as an overwhelming reality within themselves. From a broken sect they were transformed into a growing church.

They still expected the end to come soon, even during the life-time of many of them. This hope slowly withered away. Instead they conquered the Roman Empire from within in the span of three centuries. This success brought Christians into a dilemma, of which the earliest church had had no idea and which, judging from the extant sources, does not seem to have been foreseen by Jesus himself. This dilemma has a conceptual as well as a practical side. We will be concerned mainly with the conceptual side. But since it can be made clear more easily in the practical field I shall first treat it there.

Consider any practical participation in what we call today political responsibility. Serving as a soldier in the Roman army may be a good example. The first Christians would certainly have considered this to be impossible for them. It is true that Jesus himself had shown great liberality toward the publicans who served the same state, and had accepted military obedience without any hesitation as an example for faith in his discourse with the Roman centurion. Still, killing, which belongs to the office of the soldier, must have seemed scarcely compatible with Christian love, or to put it more positively, with the Sermon on the Mount. Early Christians were not pacifists in a modern sense; they did not think of eliminating war by inducing everybody to live a peaceful life. But this was because they saw history as a battle between God and Satan that would come to an end soon, not by the efforts of man but by the power of God. The single Christian had to save his and his friends' souls, not to save the world. According to the predictions of the New Testament, especially of the Revelation of St. John, they even expected that the last act of human history would be the coming of Antichrist, a human ruler who would mock Christ by establishing an outward peace which would be the absolute dominion of evil. In this way they were aware of the ambivalence of the concept of peace, and for a long time the Roman Empire could be regarded by them as the impending fulfillment of this prediction, as what Babylon had been to the Jews. But if for a long time no Christian became a soldier it began to happen that soldiers became Christians. What should they do? Especially in the troubled 3rd century, when the Roman Empire began to decline, it could be argued that the soldier was defending peace and order and a chance to lead a godly life against the barbarians. Was not his service an act of brotherly love too? Indeed, when the choice was no longer between the Empire of Rome of this day and the Kingdom of Heaven which might begin tomorrow, but between the civilized Empire of Rome and the intruding Teutonic tribes and Parthian non-believers, the aspects began to change. How much more so, after the Emperor had become a Christian.

Thus from the earliest centuries there is the tension between conservative and radical Christianity. There can be little doubt that the radical view is the earlier one. To it any attempt to help preserving the existing order of things was a compromise with the pagan gods who now were considered demons of evil—an understanding, by the way, which seems at least more profound than that of a rationalism which denies their existence. On the other hand, the conservative Christian would feel that by non-cooperation he might just have saved his conscience in too easy a way. A truly sovereign attitude, like that expressed in the Epistles of St. Paul, was probably able at all times to find positive solutions to the concrete questions of the day, which were neither narrowly conservative nor narrowly radical. But whoever has had to face decisions of this kind himself in our days knows how far we are from knowing these answers easily or in advance.

The result of Christian history under these auspices could not be anything but ambivalent. Wheat and tares were growing together. Let us first consider the wheat. We may doubt how much better the late Roman and the Byzantine Empires may have been than earlier empires by being Christian. But at least when in St. Benedict and in Pope Gregory the Great the most radical tradition of Christian life, monasticism, merged with the best Roman tradition of firm and wise government, the foundations for a new age were laid in the midst of the breakdown of antiquity. From those days the church had the responsibility for feeding the poor. In order to be equal to this task it had to cultivate rotten and virgin lands. It had to pacify barbaric rulers. Bishops became preservers of stable government. Monks carried the tradition of ancient culture through the dark ages. What we call Europe is the making of Christianity.

But a price had to be paid for this achievement. Did bishops ruling a land convert the state or pervert Christianity? Were not pride, riches, and violence the attributes of Christian rulers as they had been of pagan lords? This is human, after all; and their faults were often tempered by the understanding of every Christian that love of the neighbour was the real criterion by which human behaviour was to be measured, and that the king was not divine but under God. But was not this acceptance of a tempered human nature just an act of treason against the superhuman hope opened by Christ to the human race? Was it really true and tested, that violence could not be avoided and that the church had to be rich; could pride be a good thing under any circumstances? The expectation of the last judgment was kept alive by comparing the teaching of the gospel with the world within which Europeans actually lived.

Let us leave the political scene for a while and turn to the development of concepts, of Christian thought.

In earliest times, when the second coming of Christ was expected from day to day, no consistent self-interpretation of Christianity may have been needed. The gospel told itself from mouth to mouth, from heart to heart. But even this is an abstraction. The gospel was told in a language which had been there before the gospel, and there is no language that would not carry with it an implicit philosophy. Our first written document, the New Testament, is full of interpretation. Probably the need of explicit interpretation, indicating the new awareness of this problem, was the reason why the New Testament was written. Matthew is a Jewish interpretation, John not far from a Gnostic one, and in the Acts and in Paul's Epistles we see how a Hellenistic interpretation was arising. As Christianity was discussed and accepted by people belonging to the educated upper class of the Empire, interpretation in terms coined by Greek philosophy became its destiny. The most important school of philosophy in later antiquity was Neo-Platonism, which perhaps may be called a re-interpretation of Platonism in a religious mood and with concepts influenced by Hellenistic, mainly Stoic thought. Thus Christian philosophy or theology—the two were not yet separated—was of a Neo-Platonic character during its first thousand years. Is this fact not also subject to the parable of the wheat and the tares? Yet: how are we to distinguish wheat and tares?

In those times the Timaeus came to be considered Plato's most important work. Plato would certainly not have thought it to be so himself. He wrote many books about questions which in his view admitted of a certain answer or which were important for leading a life worthy of human beings: on the theory of the forms, on ethics, on political order. A cosmogony he wrote just once in his life-time and he called it no more than a likely story; he had to show that at least he was able to do better in this field, explaining the world by reason, than the Atomists had done explaining it by blind chance. But now the theological setting of this myth came to be considered as some premonition of the truth revealed in the holy scriptures of the Christians, while the science to which his epistemology refers was no longer understood, and his model of a city-state governed by philosophers was—as a city-state—as far from political reality of the times, as it was—being of this world—from the eschatological hopes. In the early Middle Ages Plato was transformed into some kind of a Greek Moses; and Moses, in an equally unhistorical manner, was thought to have been the author of Genesis.

I want at least to mention the views about creation, held by the, probably, greatest Christian thinker of that first millennium, St. Augustine, as he expounds them in the three last books of his Confessions and elsewhere. It is characteristic of the interpretative character of this Christian philosophy that even a thinker as original and as impetuous as St. Augustine offers his views about this subject in an interpretation of a given text which here is the first chapter of Genesis. This interpretation of course lacks all the considerations about the real history of the text which are so characteristic of Biblical scholarship since the 19th century and of which I have made use in my third lecture. To St. Augustine the text of Genesis 1 is the word of God, dictated to Moses. Much of the interpretation is allegorical, much of it is devoted to his own profound philosophical thought, e.g. to his completely fresh and original analysis of the concept of time. As an example of the allegorical interpretation I should mention that the Firmament of Heaven is understood to indicate the firm authority of Scripture and the lights in Heaven to indicate the saints. To our understanding this is very far from the clear and simple meaning of the original text; in a philosophy, however, in which every created thing somehow symbolizes God's creative will as, to use Platonic language, the copy symbolizes the archetype, allegory must seem a natural way of interpretation.

Some of the speculative statements made by St. Augustine are that God created the world out of nothing and that he did not create the world in time but that he created time with the world.

The assertion of a creation out of nothing is the final stage of the development of thought which I described in the lecture on the Old Testament, asserting God's omnipotence. A pre-existent material of the world would seem not to have been the work of God, hence somehow not subject to him. Here a clear stand is taken against Plato and for the rational consequences of the Bible. The origin of evil, in St. Augustine's philosophy, is not in pre-existent matter but in the free decision of created souls. How this combines with his views on predestination is a question that would lead us too far from our present topic.

The other statement, that God did not make the world in time, is one of St. Augustine's own profound speculations. If the world has had a beginning, contrary to the prevailing view of Greek philosophy, what did God do before the creation? Why did he create the world in the moment in which he actually did, rather than in any other moment? Such were the questions by which a philosophically trained pagan might try to expose the naive mythology of the Christian belief in creation. St. Augustine considers worth quoting, although he rejects it, the malicious answer: "He made hells for those who pry into mysteries." In fact, there is no answer to the question; we must understand that the question itself is wrongly put. God's own existence is not in time, his existence is absolute presence, not admitting of concepts like past and future. In this connection St. Augustine analyses flowing time as human beings know it, more thoroughly and in a far more modern-looking way than any ancient philosopher. But this analysis is used to show that this concept of time does not apply to God. God is not in time but he made time when he made the world. Even the expressions "he made" and "when he made" give the human aspect of creation, the aspect in flowing time, and hence not God's aspect.

We have still to consider the transformation of Plato's theory of forms in this Christian philosophy. The Form of the Good was the highest element in the hierarchy of Plato's thought. Sensual things are not truly being. The forms are truly being. The form of forms is even beyond being, as he explicitly says. This is the origin of the later concept of transcendence. The forms are the archetypes according to which the demiourgos makes the sensible world; he has not made the forms, he only sees them as they are and ever have been. Now the God of the Christians is on the highest level. A person now takes the place of the impersonal form of the Good, and the divine artisan is re-interpreted as being identical with him. What, then, are the forms? They are God's creative ideas according to which he made the world. The realm of ideas is God's infinite intellect. This is the beginning of the transformation which ended in the modern subjective use of the word 'idea'. Plato's idea is sheer or, as we would say 'objective', existence. St. Augustine's idea is God's thought. Man, made in God's image, can think God's thoughts once more, though in a finite manner. Thus the idea is rightly understood to be in the mind of human persons too. In modern times this last use of the word idea is the only thing that is left. In Locke's view the idea still somehow represents things as they are. But when this is no longer confirmed by God's ideas, even the relation of ideas to things becomes doubtful, and in the end even the meaning of the view that there is to be such a relation. This question is to be treated in the second series of lectures.

Let us turn to the beginning of the second millennium of Christian thought. In the 13th century the philosophy of Aristotle was received by Christian thinkers, being delivered to them by the Arabs and the Jews. The theology and philosophy of St. Thomas Aquinas is the most famous attempt to harmonize Aristotle with Christianity. With respect to the ideas on creation not much seems to have been changed thereby. Still, St. Thomas had to state explicitly that the creation of the world in time—as we say—is a revealed truth, not a truth that can be proved by the natural light of reason. In fact if reason is embodied in the philosophy of Aristotle this is clear, since Aristotle states that the world is everlasting.

What is interesting to us in these statements is not so much their contents as the distinctions by which they are expressed. The natural light of reason is distinguished from revelation, philosophy is distinguished from theology. These distinctions which come so naturally to a person brought up in the occidental tradition of thought are not at all obvious. They are in fact a self-explanation of a culture which had to interpret the truth belonging to one tradition of thought by the conceptual means belonging to another tradition. If we look at the actual use of the words we shall find 'reason' to mean Greek philosophy and 'revelation' to mean the Bible. Philosophy is what is acceptable of Greek thought to Christians; theology is an interpretation of Biblical thought. The distinction of reason and revelation is present already in the Fathers, i.e. in those Christian thinkers who still belong to the ancient civilization. The distinction of theology and philosophy has developed in the rising culture of medieval Europe; it is particularly apt to reconcile Aristotle with Christianity.

What does the advent of Aristotelianism mean? Look at Raphael's School of Athens. Plato's lifted forefinger points to Heaven; Aristotle's outspread hand leads our view towards the earth. This is not a precise description of the true nature of the two great Greeks, but it gives an accurate picture of what the Middle Ages thought of them. Aristotelianism means reason instead of ecstasy; it means a positive, though relative valuation of the senses instead of their rejection. The advent of Aristotelianism announces that an age of science is to come. This historical fact has been completely distorted by the anti-Aristotelian self-interpretation of the next great step towards science which was taken in the 17th century. As I said before, the tradition of Christian thought has been interpretative. This was partly due to the prominence of a Sacred Book, partly to the fact that for centuries the newly-formed European nations, rather like school boys, had to learn a great deal from ancient cultural traditions. Thus even the daring attempt towards their own ways of thought which goes on through the high Middle Ages had to assert itself by establishing new authorities rather than by rejecting all authority.

The last question I am going to ask in this lecture is: what is the meaning or the origin of this new interest taken in reality? In so speaking I already use modern language. More correctly I should say: the new interpretation of the meaning of reality? To Christian Neo-Platonism God had been the one great reality. St. Augustine is concerned about God and the soul, nothing else matters. Nothing else? he asks himself. Nothing else, is the answer. The importance attached to a thing and the reality ascribed to it go close together. To the moderate conceptual realism of St. Thomas the form of a thing, that is its participation in the universal, is its true reality, yet it is concrete only in consisting of form and matter. To later nominalism only the individual thing has real existence. This has become the modern concept of reality. For the final part of this lecture I shall use this concept of reality in a terminological manner: Modern man is concerned about reality; modern language honours a man by calling him a realist.

The average modern self-interpretation is that this concept of reality has been wrought from religious prejudice. This makes modern times appear like a new civilization, different from the other-worldly civilization of the Middle Ages. Since giving up the Christian tradition is not acceptable to everybody, Christianity is then re-interpreted as some moral code of a very high standard; this is a form in which it seems to refer to reality in the modern sense. Certainly this understanding reveals important aspects of Christianity; still I think it is a misinterpretation of the inner dynamics of the Christian faith. In the field of theoretical thought I shall treat of this question in the following lectures. I want to make clear its practical background in the final remarks of this lecture.

Let us return once more to the ambivalence of Christian history. As I said, history was steadily and incessantly transformed by those who waited for nothing but for its end. Let us follow this transformation through the Middle Ages and the beginnings of modern times. Bishops had become preservers of stable government. This was a great success of the Christian attack on unbridled nature. But taking part in government they were in a situation similar to that of feudal barons; for a long time they had to come from aristocratic families themselves. They served the king against disorder; they were even invested by the king or the emperor. But they did not belong to the Church? The Pope as the head of the Church, it was felt, ought to have invested them. And is noble birth a necessary attribute of him who is to love his brethren as the carpenter's son had done? Thus what was acceptable to one century seemed inadmissible to the next one. Christianity means a steady movement of aggression against the prevailing state of affairs.

The Middle Ages can be understood as being a sequence of Christian reform movements like that of Cluny and that of the Franciscans. The Pope won his century-long fight against the Emperor. But the Emperor-like Pope then succumbed to what was probably an alliance of spirituality in the Christian sense and reality in the modern sense. The medieval church had tried to establish an order that was spiritual in its origin and of this world in its effect. When nearly established it was no longer convincing spiritually to those who tried to be radical Christians; at the same time it seemed insupportable to the smaller rulers of this world. The Pope had already overcome the Emperor because of a similar alliance; the feeling of many good Christians that the Pope had higher spiritual legitimacy than the Emperor to establish a Christian order in this world was mingled with the wishes of feudal barons in Germany and of modern-minded city-states in Italy to establish their own power. When the French king carried the Pope to Avignon the universal aspirations of the Pope had become as dubious as those of the Emperor had been a 100 years earlier.

I shall try to describe this threefold battle by means of three abstract concepts: nature, Christianity, and reality. I do not use these concepts now in their full and ordinary meaning; I use them to describe forces active in history in words they themselves used. In the time of which I speak the feudal baron may represent nature most clearly, the Franciscan friar may stand for Christianity, and the merchant citizen of a city, say an early Medici of Florence, may stand for reality. I am not certain whether I would be able to define these concepts very clearly, although I feel those who know history will have an intuition of their meaning.

'Nature' is intended to indicate those established patterns of life and those forces in the human mind which were there before Christianity came to them. The word 'nature' is taken from the Christian vocabulary; it means natural man who needs salvation. Nature would express itself largely in the language of myth. I think this concept applies well to the early feudal society of the European nations, while its application to the old civilization of pre-Christian antiquity would be unsatisfactory.

Christianity I have tried to explain in this lecture. It understood itself as representing a supernatural power; in fact it attacked the ways of life of nature. In order that charity should act in a person the powers which naturally dominate the soul must be overcome; they must, in fact, be slain like the early gods of the myths. Nobody likes to love his neighbour, not as we think he ought to be but as he really is. This is what the fourth gospel indicates by the words "except a corn of wheat fall into the ground and die, it abideth alone; but if it die, it bringeth forth much fruit" (*John 12. 24*). God delivers us from the hands of the gods.

Reality, or more clearly secularized reality, is a word by which I try to describe the world which is delivered from the gods without belonging to the God of charity. It is the world of the autonomy of man. The hypothesis about our history which I want to bring forth here is that reality in this sense has only been made possible by Christianity. I first expressed this view when I said in die third lecture: God himself has deprived the world of its divinity. In order to give a considered judgment about this hypothesis we will have to go through the next four lectures; I shall return to the question in the last lecture of this series. Now I can only try to point to the developments in history in which similar acts appear first in a Christian and then in a secular form. I shall limit these brief remarks to three phenomena: obedience, government, and revolution.

The strictest form of obedience known to most people of our days is in military discipline. The strictest form of obedience which probably has ever existed in history was in monastic orders. St. Ignatius Loyola invented the much quoted dictum that the monk should be submitted to his superior like a corpse which makes no movement of his own. What does that mean to a Christian? The superior is to be considered the representative of God to whom really all obedience is due; the eradication of one's own subjective will is the death of the corn of wheat. I think the military orders of knights brought this concept of obedience to the military field where in feudal times discipline was far from being strict. The Spanish and Prussian soldiers came from countries whose medieval history had been largely a history of knightly orders. How far monastic and military obedience really have led towards inner freedom, is one of the many questions pointing to the ambivalence of our history. We leave it here unanswered, but we must return to it in the end.

Orderly government was promoted by bishops, as I said before. Modern political history is partly the history of an increasing power of the state. What the state subdues is precisely that magnificently unbridled nature which was so strongly represented in the feudal baron. By the 16th century in England, and the 17th century in French history the king had taken over the promotion of this development; the divine right of kings then was the modern idea versus feudalism. Equality of law and functioning of the state are a heritage left to democracy by absolutism. Later the abstract concepts of state and government take over the role of the king. The modern democratic state is even more powerful in executing its intentions than absolute monarchy was, and precisely for this reason the legal impediments against misuse of power are so important today.

But if government may in this sense be radical when in process of being established, it tends to become conservative after its establishment. What I called conservative Christianity may now be described as Christianity compromising with nature. Then Christian radicalism, if understood politically, tends towards revolution. Cromwell's Puritans considered the divine right of kings un-Christian, and so did the French Revolution, using the classical Christian concepts of liberty, equality, fraternity in a secular sense. This transformation of Christian radicalism into the radicalism of reality I should like to describe by the concept of secularization, using this term in a pregnant sense.

Prof. Dr. Carl Friedrich von Weizsäcker at the European Ecumenical Assembly in Basel (1989). *Source* HEKS/Jürg Gasser; permission to use this photo was granted

Prof. Dr. Carl Friedrich von Weizsäcker with his wife (1995). © The Weizsäcker Family represented by Dr. Elisabeth Raiser who granted permission to use this photo

Chapter 5
What Is Secularization?

The main theme of the present lecture is that of secularization.[1] This will be introduced in Sect. 5.1, and in Sect. 5.2 applied to modern political revolutions as being the most important example of it. In Sect. 5.3, the ambivalence which we already discovered in the first lecture to be a mark of our contemporary faith in science will turn out to be a characteristic feature of secularization in general. I will follow it right back into early Christian history and rediscover it particularly in the Christian idea of chiliasm, the belief in a coming thousand-year reign of Christ on this earth. Section 5.4 will look for the chiliastic features in the modern belief in progress. The attempt to understand Marx will bring us to Hegel, who sketched out a basic pattern of historical understanding as a whole. This pattern, discussed in Sect. 5.5, is ambiguous for us to a special degree: on the one hand it contains ideas which even our own analysis simply cannot abandon; on the other hand, it seems to be as it were the summit of that modern Titanism of which the faith in science of our days might almost be said to be just a harmless offspring. From this background in Sect. 5.6 I attempt to understand Marx, and in Sect. 5.7 once again return to the ambivalence of the success of the modern age.

Finally, in Sect. 5.8 I take advantage of this description of secularization to reflect one last time on its significance, on the one hand with the means of Christian theology, on the other in the light of modern science. This reflection should then clarify the philosophical significance of the transition to the systematic problems of natural science, as it shows how little it is possible to make any

[1] This text was first published as Chap. 10 of the "Gifford Lectures" 1959–1961 in: C.F. von Weizsäcker, *The Relevance of Science*. (New York: Harper and Row–London: Collins, 1964). All rights of the Weizsäcker Bequest were transferred to the Carl Friedrich von Weizsäcker Society, a nonprofit society whose executive chairman, Dr. Bruno Redeker, granted permission in March 2014 to use selected texts of Carl Friedrich von Weizsäcker.

K. Raiser (ed.), *Carl Friedrich von Weizsäcker: Major Texts on Religion*,
SpringerBriefs on Pioneers in Science and Practice 24,
DOI: 10.1007/978-3-319-03704-2_5,

historical position, even if it be the position of the very science in which we ourselves believe, the unassailable starting point of all speculation.

5.1 The Concept of Secularization

Let us first remind ourselves of the main stages in the course of these lectures up till now.

I began with the question 'What does science mean for our time?' In connection with this question, I formulated two theses, which I now repeat for reconsideration:

1. Faith in science plays the role of the dominant religion of our time.
2. The relevance of science for our time can, at least today, only be evaluated in concepts which express an ambiguity.

I then proposed, as a limited contribution to the diagnosis of our time, a study of the historical origin of this faith in science. This was to centre on a special example, for which I selected the history of the concepts of creation and cosmogony.

First, I spoke of mythical cosmogonies which were at the same time theogonies. I attempted to show how they united two elements which seem to us to be essentially different: a well-considered narrative of the physical origin of all things and the expression of an understanding of human existence. As the gods changed, so the two aspects of cosmogony changed. The God of the Jews taught his people to make a sharp distinction between good and evil, that is between life and death. Greek philosophy made a distinction between true and false, that is between being and non-being. I attempted to show how the Jewish and Greek ideas of the beginning of the world corresponded to the understanding of life and truth among both peoples. Then Christianity transformed pre-Christian nature: but the world which it had built was again transformed by modern reality. In modern times, scientific research took the place of an interpretation of the world by traditional symbols. I pursued the growth of a scientific cosmogony. This cosmogony ends, if it is expressed with scientific prudence, in open questions. This was evident in the case of the development of life, which we shall not really understand until we can give a better answer to the two questions, "What is life?" and "What is physics?" It was evident in the sphere of astronomy, where the infinity of space and time itself became an unsolved question. At the same time, however, we discovered that many people in our time hold quite firm convictions about these open questions, and we had occasion to assume that these convictions of theirs had their roots in their specific understanding of human existence. Precisely in its unresolved questions, cosmogony is evidently still a symbol of the way in which we understand the basic problems of human life. One of these interpretations is faith in science. Have we understood it better at the end of our course than we did before?

In the first lecture, I narrowed the scope of questions to make it more tractable. Now I must take the opposite step of making it more general. I shall no longer be asking about cosmogony, but about that of which it is merely a symbol. I must of necessity go about this generalizing without attempting a proof. History is too complicated to allow the strict proof of general assertions. I venture this generalizing as a diagnostic hypothesis.

I take up the two initial theses and replace them with more special ones:

1. The modern world can largely be understood as the result of a secularization of Christianity.
2. Secularization is an ambiguous word which describes an ambivalent process.

First of all, I must explain the words used in these sentences.

The word 'secularization' derives from the Latin word saeculum, which means century. In traditional Christian language, saeculum means the time in which we are actually living today, as opposed to God's eternity: hence it also means everything which belongs to this world and which to that extent does not belong directly to God. Secularization was for a long time a juristic concept which designated the transference of ecclesiastical goods into secular hands. Thus men talked of a secularized monastery. In our century, many authors have begun to use the word secularization in a more general way as a description of the process by which the modern world has developed. This use of the word implies certain conceptions which I will discuss in detail; I will begin with some negative remarks.

If the modern world is the product of a process of secularization, it is not a religious world in the strict sense of the word. It is neither properly a Christian world nor is it the world of a new religion, replacing Christianity. It would, however, be equally impossible to describe it as a world totally devoid of any relationship with a religious world. A secularized monastery is the same building as before; its rooms still have the structure of monastic cells, a refectory and a chapel, even if they are now used for other purposes. Similarly, the modern world still has the structure of a Christian world; the drawing of the picture is as it were still Christian, even if all the colors have changed, even if black has changed into white and white into black, as in a photographic negative. If this is so, however, the process can only be ambivalent and the concepts in which we express it must necessarily have an ambiguous sound. For the problem is: Are we to stress the Christian structure, or the non-Christian use of it?

After this explanation of the terminology, I will give examples for the new theses. One example is the origin of modern science, as we encountered it, for example, with Galileo. In the sixth lecture, I asserted that the concept of strict and generally valid laws of nature could hardly have arisen without the Christian concept of creation. Matter in the Platonic sense, which must be "prevailed upon" by reason, will not obey mathematical laws exactly: matter which God has created from nothing may well strictly follow the rules which its Creator has laid down for it. In this sense I called modern science a legacy, I might even have said, a child of

Christianity. But then I had to show how science lost contact with its parental home. Children can experience the death of their parents.

We have already seen in Galileo's conflict with the Church the ambiguity of any concept which describes this process of secularization. Was Galileo right, when he read the greatness of God in the book of nature, in thinking that he was fulfilling God's will that men should read this book? Was the Church right in thinking that this would distract men from the will of God which stands written in the book of redemption? Using the categories of our century we could describe the positions as being equally ambivalent. If we leave aside the violent means which the Church used wrongly, and, in the long run, used unsuccessfully precisely because they were wrong—did the Church want to hinder the progress of knowledge, or did it require a wider field of vision than that of the fanatical specialist?

The concept of infinity offers another example. The majority of pre-Christian world-views knew only of a finite world. For Christian philosophy, the world was similarly finite, but God was infinite. In modern times, the world takes over this attribute of God: infinity becomes secularized. Under this aspect it is most remarkable that our century has begun to doubt the infinity of the world. I believe that in our time a critical examination of secularization is beginning at exactly the same time as secularization is achieving a consistency hitherto unknown.

5.2 The Political Revolutions

In order to see the full weight of secularization, however, we should not limit ourselves to theories. Instead we should speak, for example, of politics. I therefore take up the thread where I let it go at the end of the fifth lecture. There I pointed to the Christian background of three modern phenomena: military obedience, ordered government and political revolution. Allow me now to develop the last of these three themes further.

At first glance, the political revolutions of Europe display many differences, and that should not surprise us, as history never repeats itself. I have spoken of the English Puritan revolution of the 17th century and of the French grande revolution of the 18th century. One difference is that the first of the two interpreted its aims in strictly Christian terms, whereas the second even went through a stage of militant opposition to Christianity. These two revolutions were not notably successful: the revolutionary governments collapsed, and in each case an age of restoration followed. As movements akin to them, which were at first sight more successful, I should mention the American revolution of the 18th century and the Russian revolution of the 20th: both erected systems of government which have lasted down to our time. These systems of government again differ from each other; anyone asserting that one was like the other in any essential feature would in fact win little political sympathy either in the western or in the eastern world. Despite everything, however, I regard it as historically true that all these revolutions, and similarly all the political systems which they set up, either immediately, or after several apparently abortive beginnings, have a great deal in common with each other, and I hope to trace just this common modern element in them. It is the ambivalence intrinsic to modern civilizations which in my view makes it easy for these systems to regard themselves as being so different: each of them allows of opposed interpretations, each of which contains a good deal of truth.

In speaking in this way, I must, however, make my own standpoint quite clear. I belong to the western world. I share the western conception of political freedom and the rights of man. My remarks would be misunderstood were one to find in them a tendency to obliterate differences in approach to questions in which right and wrong is at stake. A decision is necessary, even if we see what is false on our side and what is true on the opposite side. But I fear that we squander the good consequences of such a decision if we allow the necessary decision, once made, to make us blind to the common feature of all modern systems. We are directed to live with one another in one world. Just when we rightly and passionately hope that systems hostile to our own will develop to the standards of humanity which we hold to be the only permissible ones, just at this point, we may not identify these standards with our own historical prejudices and with our own highly ambiguous actions. If self-control is a prerequisite of any ordered conduct in human life, those at least who possess the necessary gifts of understanding and factual knowledge ought to exercise this intellectual self-control in their views about the side which they have chosen as their own in the present historical struggle. This attitude is not only a requisite of good taste: there could come moments in which survival depended on our being capable of it.

I am thus putting forward the supposition that revolutions have very much in common. Let me take the slogans of the French Revolution as a starting point: Freedom, Equality, and Brotherhood. If I see it rightly, these words express a common aspiration of all these different revolutionary movements. The distribution of emphasis may have varied. Perhaps Cromwell's Ironsides tried above all to obtrude their understanding of brotherhood, Robespierre's Jacobins their understanding of freedom, and Lenin's Bolsheviks their understanding of equality on their fellow men. But as soon as we take the three concepts as seriously as they deserve, they do not allow of separation. Can we speak of true freedom as long as it rests on the servitude of a part of society? If not, freedom also demands equality. Can we maintain equality by brute force? If not, equality rests on brotherhood. Can I honestly call my neighbor my brother if I do not accord him the freedom I claim for myself? If not, brotherhood demands that a fellow man be permitted his freedom.

5.3 The Christian Background to the Modern Ambivalence

The brief analysis of the three basic revolutionary concepts which I have just given was, I think, a Christian analysis. In this sense all modern revolutions have sought to realize Christian demands. But why have they then so forgotten their Christian background that in the most recent of these revolutions it has been felt impossible for the same man to be both a Communist and a Christian at the same time?

If I see things correctly, this progressive amnesia on the part of revolutions corresponds to a quite similar forgetfulness on the part of official Christianity. In the fifth lecture, I spoke of the intrinsic ambivalence of Christian history itself, and sought to describe it by means of the concepts of radical and conservative Christianity. The earliest Church was radical, but not in a political sense. As the success of Christianity brought political life within the purview of Christians, there gradually developed the medieval Church, which in a comprehensive, a catholic way embraced both the radical and the conservative elements. This Catholicism did not mean peace: it meant the incessant strife of the opposed tendencies within the Church, and this strife means life. In modern times the strife has torn apart the catholic unity. The Reformation divided the Church into disputing churches, and secularization divided the world into an official Christianity which tends towards pure conservatism and a non-Christian world whose radicalism no longer understands itself in the light of the Gospel.

If we consider in detail the steps which have led here, today's conditions must seem to be the natural consequence of a necessary development. It is to make that quite clear that I have devoted so many lectures to an extremely specialized problem. But if we look back from our present position to the beginnings of Christianity, the result must seem paradoxical, self-contradictory to the point of absurdity. Let me analyze these self-contradictions.

The Church preserves this most revolutionary document of human history, the Gospel, the truth of which is slowly slipping from the consciousness of the citizen of our modern world. As the Church knows that what it preserves is the truth, it allows itself to be led into the attitude which is assumed by all who preserve the supreme good against the changing trends of the day: it allows itself to be seduced into conservatism. At least, we will hand down uncorrupted to future generations the good, with which we have been entrusted! Laudable as this conservative attitude is, the facts of life force it towards a position which is not so very different from the equally laudable and equally inadequate position of the Scribes and Pharisees. And as usual this is seen much more easily from outside than from within. The Christian concept of the Pharisee, applied to the Church itself, might easily be the last Christian concept the modern world forgets.

Even the person who sees this from the inside cannot ipso facto alter it. Despite many still wonderful works and achievements of Christians throughout the world, and despite some hectic exertions on the part of Church officials to keep pace with modern times, the Church has for several centuries no longer been leading the historical process; it can hardly still follow it. The most profound thought possible within Christian conservatism is therefore probably that the Church should not in any way lead the process of history; indeed it should not even follow it, as this process is self-destructive, or at least alien to the Gospel. Where this voice has a genuine ring we may not ignore it. But I fear that even it expresses only one side of the truth, and not the complete truth. I feel that it too remains within the sphere of ambivalence, and the consequence of an undetected ambivalence in one's own attitude is blindness to the facts.

The modern world, however, is no less blind. Take one more look at the revolution. La revolution devore ses enfants. We know how blind violent insurgents are to what they really achieve. Did the Ironsides bring brotherhood? Did the Jacobins bring freedom? Did the Bolsheviks bring equality? I give no answer to these questions, not even a negative one, as even this straight-forwardness, judged by the later course of history, would probably be an exaggeration. I would, however, ask: Why this ambivalence of revolutions? Has it perhaps something to do with an age-old dilemma, well known from Christian history, that of violence and non-violence?

The aim of the three revolutions of which I have just spoken was a state of society in which rule by force would no longer be necessary, whether this state of society was given the name of the Fifth Monarchy, the Age of Reason or the classless society. But the way taken by the revolution to this supersession of force is, in fact, force. Of course the thoughtful leaders of the revolution will argue that the ruling power can never be overthrown except by force. Here the leaders of the revolution are expressing a well-known conservative view of human nature. In all previous history those men have ruled who were ready and able to defend their rule by force. It is perhaps the most revolutionary idea of revolutions that this need not always be so. This idea derives from Christian eschatology. But is the revolution that fights for this idea the sole power in history that is justified in using force? Or does it thereby sacrifice its aim to the devilry of the means which it regards as necessary?

This dilemma of revolutions seems to me to throw a special light on the thesis of the mutual blindness of Christianity and the modern world. Christianity does not recognize its own concern in revolutions and is thrown on a fruitless defensive. But revolution sees Christianity only as the guardian of what must perish and therefore forfeits the possibility of understanding its own concern with the aid of concepts which reach deeper than those which it can itself offer. I am therefore attempting once again to portray the dilemma here as it has revealed itself in Christian history—which must be regarded as pre-history to modern revolutions.

In Christianity, the relationship between the end and the means is less simple than in its pagan precursors. In Christian terms, paganism divinized human nature and along with it its intrinsic tendency to use force: this can be seen even in so spiritual a creation as Plato's model state and in a form of government as rooted in ethics as that of the Stoic emperors of the 2nd century. Christ completely rejected the divinization of our natural habits. In the coming kingdom of heaven the rule of these demons will be broken. But, asked the following Christian era, how will the kingdom of God come? It does not come through our efforts: it comes of its own power. This thought contains a deep insight into human existence, which finds its application in daily life. If the demons are conquered, they do not succumb to our efforts of will, but to an operation which we can experience only as a gift of grace. This, however, does not in the least dispense us from any exertion of our will at all. Grace is the answer to our longing for grace, and this longing is not in earnest unless it leads to the most determined personal effort. It is a well-known doctrine that man cannot give himself grace but that he can squander the grace that is offered. This is in accord with the teaching of the Fourth Gospel, that judgment takes place in this life. Should things be otherwise in the great course of world history? But what would this mean for world history?

The final victory over the rule of force is the Last Judgment, the Second Coming of Christ. The orthodox Christian teaching was that no human action could influence the time of his Second Corning. In the present span of time, in which we live in this expectation, we should attempt to live as citizens of his coming kingdom. How this was to be done in practice was, however, the question which produced the tension between radical and conservative Christianity, and I expressed the view that this very tension was the driving force of world history in the Christian era.

This tension has at times expressed itself in a subtle point of dispute within the Christian theology of history. The Revelation of St. John speaks of a thousand-year kingdom of Christ and his saints before a new heaven and a new earth are finally made. What did this millennium represent? The present era of the Christian Church? Could this confused time be regarded as a rule of Christ and his saints? Or was the thousand-year kingdom yet to come, perhaps soon, perhaps still within the lifetime of the present generation? Did its coming perhaps still depend on our acting rightly? This belief, that it would come soon and that we ought to act accordingly, was called chiliasm (chilia ete = a thousand years); the Church regarded it as heresy, but many social revolutionaries of the late Middle Ages and early modern times took it over. But chiliasm itself revealed the dilemma: How then should we act to help bringing in the kingdom? Will the rule of non-violence be brought about by non-violence or by violence?

Only a few have at any time attempted really to live according to the Sermon on the Mount. For a long time, the recognized form of this attempt was the life of the monk. In a characteristic fashion, the great prophet of chiliasm in the early 13th century, Joachim of Fiore, taught that there were three Churches following one after the other: the Church of the patriarchs in the Old Testament, founded on Abraham; the Church of the priests from the time of Christ's first coming, founded on Peter, and the Church of the monks, now beginning, founded on John. Alien as this way of speaking is to us today, it is the spiritual tradition in which the European revolutions still live. Joachim expected the coming of the new kingdom through God's rule over history; but man had to live according to this kingdom if he and the brothers who would be reached by his voice and his example were to share in it and not fall into damnation. For this hope, the 'Spirituals' of the Franciscan order, among others, deeply influenced by Joachim, made any sacrifice, including that of their lives. Later, in the Protestant world, the natural form of the attempt to live strictly after Christ's commandment was the sect. The most splendid example of what a life lived according to the unqualified Sermon on the Mount can achieve, and one that is most visible to modern eyes, is that of the Quakers. The Church was indeed often right in the face of the exalted expectations and fantastic scriptural interpretations of the 'enthusiasts', but any glance at the past or the present will show that 'enthusiast' is also an eminently convenient formula for doing down those who attempt to take Christianity seriously.

When I said that the European revolutions still lived in the spiritual tradition of chiliasm, I did not primarily have to think of those who renounced a violent contribution to the coming of the kingdom, but of the chiliastic social revolutionaries. Important as this distinction is, however, belief in an imminent radical change in world history is common to both sides. A secularized form of this belief governs the stormy course of modern history.

5.4 Belief in Progress

The one tenet of faith which the western world today shares with Communism—and, moreover, with the nationalist movements of non-European countries—is a belief in progress. This is a recent idea in history. Neither European antiquity nor the great cultures of the East understand history as a field of progress. For the Christian Church, there was only a progress in history which had already taken place, the first coming of Christ, and a second progress in the future, the end of history in his coming again. But chiliasm allowed a transposition of this otherworldly salvation of history into the history in the midst of which we live. From a historical point of view, I regard belief in progress as a secularized chiliasm. Of course, modern man would understand chiliasm in the opposite way, as a dawn of the belief in progress, still verging on the mythical. Be this as it may, only the religions of the Judaeo-Christian tradition seem to have the concept of an unrepeatable history of our world, and I would not regard it as fortuitous that the modern concept of history has arisen in a Christian culture. Were I to return to the philosophical problems, I would at this point have to investigate the Christian and the modern concepts of time. I hope I might be able to make it probable that our concept of history is not less, but to a still greater degree, a legacy of Christianity to the modern world than is our concept of the law of nature. But that would take me too far afield today. I will now investigate the particular understanding of progress in various modern political systems.

The western world understands its own scale of values to be one which preserves what was good in the Christian tradition; it therefore finds little difficulty in linking its belief in progress with Christian conceptions. The central political idea of the West is the freedom of the individual, and without doubt Christianity has taught us to take any living human being seriously as a person, no matter to what class, race or nation he or she belongs. Liberalism might well claim for itself that it took Christian convictions seriously in advancing this respect for the person in the political sphere as well, often against the resistance of conservative Christians. Even technical progress can, at least by way of the natural science depicted in these lectures, be linked with Christianity: technical science can understand its work as the fulfillment of the commandment "subdue the earth under you". The problems of the modern world, with which I began in the first lecture, can then well be understood as a consequence of the ambiguity of a secularized Christian world. But this thought itself contains an ambiguity which I must investigate in detail at the end. Before this, however, I must turn to the apparently expressly anti-religious doctrines of Marxism. If we are to do them justice, we must allow ourselves to be led on to a new plane of historical understanding.

I would express and advocate the by no means original view that in some very important points Marx was nearer to Christianity than the committed Christians against whom he turned. I can only develop this view in two stages. Marx is only to be understood as one who trod a way opened by Hegel. I must therefore turn first to Hegel, as much for his own sake as for his significance for Marx and Marxism.

5.5 Hegel

I am not concerned here with Hegel's perhaps all too artificial and untenable particular theses. The time in which we had to fight our way free of Hegel's influence, natural scientists, philosophers and theologians alike, is past. I should like to show which way of thought was only made possible by Hegel's approach to the question of philosophy.

One can call Hegel the first philosopher who regarded history as a philosophical problem. The central theoretical problem can easily be elucidated by a retrospective glance at these lectures. When I spoke of views of the origin of the world, I had always two simultaneous questions in mind: Why did men believe this or that in this or that phase of human history? and, Which of these views has shown itself to be true? If we take both questions equally seriously, we are brought into a confusing situation. We ourselves live in a particular phase of history. My present theme, that of secularization, may be described by the question "Why has our own age the views which it in fact had?" But if I thus make our own views historically comprehensible, where is our naive faith in the truth of these views? And if I make a radical reduction of truth to historical relativities, what do I mean when I declare my own views, e.g. just this historical relativism, to be true? It is indeed so easy to show how well historical relativism fits the thought-pattern of modern times; i.e. it is so easy to explain it itself in the frame of historical relativity. In this morass of historical relativities, where can we still take a firm step?

Hegel was the first thinker to understand this question, and he sketched an answer. Perhaps I may indicate this answer in simplified expressions, beginning by leaving aside an explicit use of Hegel's central concept of the Absolute. According to Hegel, truth is not real otherwise than in the form of historical positions following one another: each of these positions is to some extent the truth. But it is only the truth to some limited extent. The dialectical process of history is maintained by each position showing itself to be one-sided once it has prevailed. Therefore truth itself now demands a new, contradictory, position, which supersedes the previous one only to suffer the same fate after its own victory. In this constant change, the truth which was there in previous stages is at the same time both superseded and preserved; to use Hegel's intentionally ambiguous expression, it is "taken up[2]" in the later stages. If we keep to this quite general understanding of Hegel's, and leave on one side his artificial systematization of history (instructive as it is in individual points), we shall have described what is probably the only way that we too can follow. I have in fact constantly used this mode of thought in the previous lectures. I believe that it can also be applied consistently to our own position in history. Even this application I hint at here only in a somewhat superficial way, but one which will perhaps be sufficient to use for the moment. As far as certain ideas reveal themselves inescapably to our honest, self-critical urge towards the truth, as being true, we shall be justified in using them as a basis for our judgments. Should the future set them aside,[3] we can do nothing about it: but we may trust that the element of truth in them will be preserved. Otherwise a man who has recognized that he stands in the midst of history could hardly pass judgment on his own thought.

But this last remark compels us at one, equally well known, point to criticize Hegel, at least to criticize the picture which his philosophy presented and had to present to his time. His own thought claims to be a system, and it must be so, intending to think the Absolute. But "the absolute is the result"; Hegel did not at least stand in the way of the impression that it is Hegel's philosophy which is this absolute result of the history of the spirit. Understood in this way, his philosophy appears as the most titanic chiliasm there has ever been. The word chiliasm is in place here. Hegel understood his philosophy as the genuine understanding of Christianity: this problem of understanding Christianity was his starting point in his youth. The definitive understanding of Christianity is, however, the millennium. Thus, even in Hegel's political philosophy, the early 19th century state was not to be distinguished from the millennium with complete clarity. (It is, however, unjust to ascribe this view to him in the particular case of the Prussian state in which he lived for his last years; he was not uncritical of this state, though careful in his expressions, and one might more accurately call him a philosopher of the French Revolution.)

5.6 Marx

Here we reach the point at which Marx criticized Hegel. Marx was a real Hegelian. He believed, perhaps in the last resort more literally than Hegel himself, in a final stage of history. But this stage had not been achieved—how could people assert its presence when one thought of the fearful social conditions which the beginning of the industrial revolution brought with it? And it was to be achieved, not by further

[2] German: *aufgehoben.* The verb can mean "take up", 'preserve' or 'abolish'; it is impossible to reproduce the word-play in English.

[3] *aufheben,* see note 3.

thought, but by action. Hegel seemed to have dissolved history into a charade of spirits, called ideas. To describe the complete reversal of thought which is necessary for philosophy, he coined the somewhat mysterious maxim that the philosopher must learn to walk on his head. Marx boasted that while philosophy itself had been standing on its head in Hegel's writings, he had once again set it on its feet. The feet are Marx's economic ideas. Now whatever may have been ill-considered or false in the details of Marx's economic theory, he stressed a most important and, I would say, a most Christian truth. One of Jesus' terrifying remarks is that it is easier for a camel to go through the eye of a needle than for a rich man to enter into the kingdom of God (*Mark 10. 25*). Marx understood the truth from which this remark derives. He understood it in secularized form and may have exaggerated it, but he understood it. "Where your treasure is, there will your heart be also (*Luke 12. 34*)"; that is the irrefutable truth behind Marx's conviction that history is guided by economic conditions and that anyone who would alter the life of man must alter economic conditions. How many committed Christians understand this truth?

In the question of the appropriate means of bringing about economic and social changes, however, it seems to me that Marxism was open to the ambivalence that is my theme. Marx himself was somewhat ambiguous in this respect. On the one hand, he believed in the necessity of history; he believed in the inevitable coming of the classless society in the same way as Christians believe in the inevitable coming of the Last Judgment. Perhaps one can even say that his teaching gained so many adherents because he offered a positive, though secularized, eschatology at a time when Christians no longer believed firmly in their own eschatology and thus aroused the justifiable impression that the foundations of their faith were affected with doubt. In contrast, Marx preached action; he did not expect history to progress otherwise than through men themselves. The question of the right means of action led to the division of the International Socialist Movement into those who believed in an evolution within the framework of parliamentarian democracy and those who felt it was necessary for the existing order of society to be overthrown by force. Decisions about the means do not remain without influence on aims. Today, the gulf separating communism from western Socialism is hardly less deep than that separating it from pure capitalism. Whether these two gulfs can ever be closed if the ambivalences in the present attitudes of all sides are better realized, we cannot decide today.

5.7 The Ambivalence of Success

I leave the ambiguities of the aims and means of revolutions and turn once again to the ambivalence of their success. Earlier, I distinguished revolutions which were immediately successful from those which were not. On closer investigation, this turns out to be a superficial distinction. In one sense, all revolutions achieve their aims. The restoration of the Stuarts and the Bourbons lasted just long enough to make it clear that the past would never return. In another sense none of these revolutions achieved its aim. The final result of the English revolution was not the kingdom of God, but representative democracy. Neither Rousseau's return to nature nor the Age of Reason developed from the French and American revolutions, but the irrational dynamic of the civilization of the industrial age. Communism has so far brought into being not the classless society, but a highly successful hierarchy of functionaries and specialists. Modern competitive societies are much nearer each other in their actual conditions than in their programmes. Henry Ford's well-known view that the solution of the social question lies in increased production appears today as the practical doctrine of salvation of the Russian communists, while in capitalist countries a managerial class is tacitly suppressing independent entrepreneurs. Human life is planned to an unprecedented extent. Full employment, a free week-end, technical appliances and amusements for all are generally the sought-after ideals, industrialization and birth control the pressing problems, nuclear war, 1984 and the Brave New World the nightmares of this society. And its prevailing common faith is, as I said in the first lecture, faith in science. But this faith is deeply ambiguous. If it keeps to the truth, as according to its own basic doctrine it must, it must concede that it has not understood man's nature and that it does not know where progress will take us.

5.8 What Is Secularization?

We have come full circle; we find ourselves back at the starting point. We will attempt some final reflections on the results. I chose a strong expression for the ambiguities of our time in saying that while the Church was blind to the true nature of modern times, the modern world was equally blind to its own nature. Both are blind to the significance of secularization. I said that the modern world was the result of a secularization of Christianity. That means that the modern world in certain respects is, and in certain respects is not, a Christian world. Contrary to the beliefs of many Christians and all secularists, I tend to the view that the modern world owes its uncanny success to a great extent to its Christian background. If the men who think that Christianity rests on the deepest insight into human nature which has yet been revealed to us in history are right, this view should not surprise us. To repeat it in traditional Christian language: the gods of nature have been vanquished by the God whom Christians call Our Father; therefore man, as God's son, has received power over nature. As he is son and not servant, he is free, and his freedom includes the freedom to act against the will of his Father, the God of love. He can now subject the world to himself, and secularism does precisely this. (In these last sentences I have followed as faithfully as possible Friedrich Gogarten's theology of secularization.)

But these thoughts need expansion. We must go one stage further in their Christian interpretation and then we must consider them once again from the modern side. It may be helpful for a Christian theologian to imagine secularism as a Christian heresy. According to the theological definition, heresy (hairesis) is to take (hairein) a partial truth from the whole of the Christian truth and to make this part absolute. A heretic is a Christian, but an erring Christian. Different varieties of secularism select different aspects of Christianity; I will refrain from going into the details of these aspects. All select the truth that this world is to be changed, a truth which Jesus elucidated in his parables by such different examples as that of the grain of mustard-seed growing into a tree in which many birds can build their nests and that of the all-consuming fire. The modern world is a tree in which many birds build their nests, and it is an all-consuming fire. It is, I say once again, ambivalent. But it is heretical, i.e. blind to the other side of the truth, because it is blind to its own ambivalence. Belief in progress is a half-truth. Jesus clearly described the inevitability of ambivalence in the parable of the wheat and the tares, which grow together and will only be separated at the end. I have never seen a clearer description of modern times than this growing corn-field, in which the tares unavoidably grow up alongside the wheat. But he who sees the fact of this ambivalence has taken the first step away from it; he is forsaking the error that made him a heretic. Anyone, on the other hand, who does not see the ambiguity has fallen a hopeless victim to it.

Now I must turn the tables once again. It is not sufficient to call secularism a Christian heresy. One usually thinks of a heresy as the selection of part of a recognized totality of truth. The theologian who speaks of heresy is generally convinced that the Church possesses the all-embracing ('catholic') truth. This conception probably does not even describe the celebrated heresies of the past at all accurately. As a rule, the heretics stressed a side of Christianity which the Church had not taken seriously enough. Many of the Church's dogmatic decisions were called forth by heresies, and I think that the best dogmatic decisions of ancient times had a truly paradoxical character because they incorporated a truth brought to light by heresy in a thought-system which was using an apparently contradictory language. Be this as it may, secularization forces us to attempt a new interpretation of the Christian faith. This new interpretation has been going on for centuries, but it is by no means completed.

I have been at pains throughout these lectures to take account of this new interpretation. True, I have attempted to speak of ancient times in their own language, but I have also attempted to make it clear on each occasion that this was a language of times long gone by, a language which could perhaps express truths which many of us have forgotten or never learnt, but not a language which we moderns can honestly speak of as our own. I have treated many time-honoured religious views as myths or legends and have not attempted to conceal my agnosticism in the case of others. But my concern was not what is often called in modern theology the 'demythologizing' of Christianity. As I myself have spent the greater part of my life in surroundings influenced more by natural science than by the Church, I feel the battle of demythologizing to have been decided long ago, perhaps in Galileo's day: it only remains for us to be honest in its consequences. Science has come into being and will, to judge by human standards, endure; in face of it there remains only the task of interpreting Christianity in a way credible to a thought schooled in it. I have attempted this task just as many other modern thinkers inside and outside theology have attempted it. What really concerns me is another question.

I touched on this question in my remarks about Hegel, and I must now outline it more precisely. Modern thought expresses itself most coherently in science. But we have no occasion to take science itself as an absolute truth. This holds in every detail. It is indeed virtually the most imposing feature of science that it requires of its disciples that they should be ready at every moment to re-examine even the most generally accepted scientific doctrines. The life of true science is a life of constant self-correction. This must be true still more in respect of the scientific attitude to the wide areas of human life in which science, as we know, does not, at least today, have the answer to the burning questions. And how sure are we of the general philosophical background to science?

If the views of secularization advanced here are correct, they give us additional information which at first sight will not lessen our embarrassment. If these views are correct, modern science would not perhaps have been possible without Christianity. In that case, we are evidently moving in a circle. We explain Christianity in concepts which are to be comprehensible to scientific thought: a thought which for its part is found itself to be a product of Christianity. But any thought which has become conscious of the relevance of history must move in a circle of this sort. Today we can reach a stage of consciousness at which the historical naivety of those who identify their own standpoint with absolute truth must vanish away. Once we have understood this, we are none the worse for it, for no one will be expected to base his judgments on anything but what he can know. We now understand only our task of constant self-correction better than before. If we are philosophers, of course, we shall have a concept of truth which accords with this stage of consciousness.

A philosophy which tries to achieve what is required by this stage of consciousness will presumably have to describe the circle of the mutual dependence of our concepts more than once. Perhaps I may be allowed to remark here that in my earlier lectures on the history of nature I attempted to describe a semi-circle of this movement. There I took the concepts of natural science for granted and attempted to show how human history grows out of the history of nature. In the series of lectures now ended I have described a second semi-circle. Here I have taken for granted the concepts of human history and attempted to show how modern natural science has grown out of human history. In the following series of lectures I shall venture on the first semi-circle for the second time. There, my theme will again be scientific statements about nature. But I shall not then simply accept them: I shall analyze their significance in the light of what we have learnt in the first semi-circle. That means that we must return to completely abstract questions. I am sorry about that, as I feel, as I am sure you do, that the now so frequently mentioned human problems are the burning ones. But one may not neglect analysis, and I would not feel myself up to any further analysis of the history of our time if I had not first made a critical examination of the concepts of our science, the tools in the workshop of modern rationality.

Prof. Dr. Carl Friedrich von Weizsäcker with his wife in Japan 1974. *Source* The photo collection of the C.F. von Weizsäcker Society represented by Dr. Bruno Redeker who granted permission to use this photo. Original source unknown

Chapter 6
The Sermon on the Mount: Interpretations

6.1 Prefatory Note

In preparing the lecture on Anxiety for the church congress (see in this volume Chap. 14) and mainly in preparing the Bonhoeffer lecture (see in this volume Chap. 7), I read during the summer of 1975 parts of the New Testament in the Greek version, foremost the Epistle to the Romans and the beginning of the Sermon on the Mount. Connotations of the words in these texts struck me, which had been obscured by Luther's text, familiar to me since childhood, and basically by the church's entire tradition of exegesis.[1] My attention to these meanings had been primarily sensitized by a decade's practice of reading Greek philosophy in the source texts that had become part of my philosophical teaching responsibility. I might add that my personal encounter with the reality of Asian religion played a role as well…

When I read the Greek text of the familiar Beatitudes with this thoroughness, the scales fell from my eyes—that, in any case, was the subjective experience; I saw that I had to change my previous interpretation of the text on one essential point, a point connected with a better knowledge of the speech's audience, of its "seat in life." This may be gathered from the following translation and subsequent commentary. Normally, when reading the Greek text of the New Testament, I consult the translation of my Great-Grandfather, Carl Weizsäcker. If in the Greek text there is a word that allows for several meanings, this wise and careful,

[1] This text has been translated from Chap. IV. 5 "Die Seligpreisungen" in: C.F. von Weizsäcker: *Der Garten des Menschlichen* (München: Hanser Verlag, 1977): 488–508. All rights of the Weizsäcker Bequest were transferred to the Carl Friedrich von Weizsäcker Society, a nonprofit society whose executive chairman, Dr. Bruno Redeker, granted permission in March 2014 to use selected texts of Carl Friedrich von Weizsäcker.

K. Raiser (ed.), *Carl Friedrich von Weizsäcker: Major Texts on Religion*, SpringerBriefs on Pioneers in Science and Practice 24,
DOI: 10.1007/978-3-319-03704-2_6, © The Author(s) 2014

liberal translator usually has found a German word that has the same quality of several meanings. My present translation purposefully abstains from the virtue of my forebear; it commits itself to an interpretative decision and is therefore at places more an interpretation rather than a literal rendering. Just for that reason it does not claim to be useful in the church but invites academic discussion. The conclusion is a discussion of this text's historical influence.

6.2 Translation

Matth.3, 2: Change your whole consciousness, because the kingdom of the heavens is here.
4, 8: Again, the devil takes him along to a very high mountain.
4, 10: Thereupon Jesus says to him: Go away, Satan; because it is written: You should worship the Lord your God and serve only Him.
4, 11: Then the devil leaves him, and behold: angels came and waited on him.
5, 1 When he saw the crowds, he went up the mountain, sat down and
5, 2: [5, 1]: his disciples came to him. [5, 2]: And he opened his mouth and taught them, saying:
5, 3: Blessed are the mendicants, for the kingdom of the heavens is here for them.
5, 4: Blessed are the mourners, for they will be comforted.
5, 5: Blessed are the mild, for they will inherit the land of Israel.
5, 6: Blessed are those hungering and thirsting to be righteous, for they will be satisfied.
5, 7: Blessed are the merciful, for they will receive mercy.
5, 8: Blessed are the pure in heart, for they will see God.
5, 9: Blessed are those who bring about the kingdom of peace, for they will be called sons of God.
5, 10: Blessed are those persecuted because they are righteous, for the kingdom of the heavens is here for them.
5, 11: Blessed are you when they revile you and persecute you and tell all sorts of bad lies about you for my sake.
[5, 12]: Be glad and rejoice, because your reward will be great in the kingdom of the heavens. In the same way they persecuted the prophets before you.

6.3 Commentary...[2]

[2] The detailed exegetical commentary referring to the Greek text is not included here.

6.4 Interpretations

For 2000 years this text, the beginning of which we have read, could not come to rest. Whoever—affected afresh by it—attempts to interpret it anew, will probably soon find himself within the scheme of one of the interpretations that the past has already offered. Thus, it may contribute toward self-examination to present an overview of these interpretations. I would like to discuss six manners of reacting to the Sermon on the Mount, four Christian ones and two non-Christian ones. Christian: by the Catholic Church, Protestants, sectarians, and modernists; non-Christian: by socialists and Hindus. None of these groups has a unique, intrinsically uniform reaction, of course. What is meant is that each, according to its own principle, is disposed toward seeing a particular aspect of the Sermon on the Mount.

People for whom the Sermon on the Mount is not a traditional authority find it easier to read it impartially and directly. They don't have to be afraid to read what is written there, because, if it does not convince them, they are, of course, allowed to reject it or leave it alone. That is why the non-Christian positive assessments about Jesus and early Christianity may be the most instructive to start with. The underlying reasons for approvals by socialists and Hindus are almost directly opposed to each other.

Let us begin with *Hindus* of the past hundred years. Swami Vivekananda's appearance at the World Parliament of Religions in Chicago 1893 became famous.[3] In this case, as with the following groups, I do not attempt to present particular authors, but rather to characterize a general perception. It does not pose any obstacle to Hindus, if Christians ascribe a divine nature to Jesus, understanding him as the son of God. According to Hindu conviction, from time to time throughout the history of humankind, people appear as incarnations of God in the true sense—or of a divine person. Everything in the above account that is strange to practicing Christians does provide evidence for a Hindu that Jesus was one of the greatest divine incarnations. The incarnated person, manifesting divine presence, lives free of all bonds to possession, power or carnal love; he teaches, consoles, heals the sick; he sees God; and indeed, as a fully actualized human being, he is God and we see God in him. He lives in retreat on the mountain with a few disciples or as a wandering preacher; he teaches no system, his words in every conversation come to him spontaneously and go straight to the heart of his counterpart, and thus these words remain effective throughout history, if they are passed on. He does not change the social order; rather the only place where a change can be initiated is in the human soul. Whoever wants to follow him radically, should adopt his way of life. For every spiritual level there is an appropriate way of life, since the spiritual and the material are ultimately just aspects of the same reality. The lifestyles of the hermit and wandering monk have existed in India since thousands of years, and Jesus' every word and every

[3] Cf. *Vivekananda. The Yogas and Other Works Including the Chicago Addresses…with a Biography* by Swami Nikhilananda, New York 1925.

action show that he was familiar with the experiences of these ways of life. The Christians just made one mistake. They proclaimed Jesus to be unique, as *the* Christ and thereby relegated him to an un-emulatable and—to that extent, noncommittal—other world, even though, according to St. Paul, Christ is the first-born among many brothers (*Rom. 8, 29*). Nothing speaks against a Christian committing himself to Christ as his redeeming and authoritative Lord and master. But the intolerance that knows only Christ to be divine is, in truth, the protective wall of the deep fear by Christians of the full reality; especially of the demand that Christ poses to them personally; of a loss of reality that makes them militant, dishonest, neurotic and turns them into Pharisees.

Some radical *socialists* see Jesus entirely differently. To them he is a social revolutionary in a period that did not yet have any scientific understanding of social conditions. They can produce some credentials for their view. It certainly does appear that Jesus always sided with the poor and socially oppressed. Some of his exclamations anticipate Marxist views. Sayings, such as, "*For where your treasure is, there will your heart be also*" (*Matth. 6, 21*) express the psychologically defining role of economic reality, which was being denied by Christian capitalists. Hindus know this too. But unlike the great Hindu teachers, Jesus attracted political prosecution. Irrespective of the question whether the motivations for the prosecution and condemnation may have been influenced by a misunderstanding about his teachings, there is ample evidence that he did not conceive the kingdom of the heavens as a sectarian community but as a transformation of the whole world. His thinking was oriented towards history as was the case for prophetic and Messianic Judaism. And the primitive Christian community to a large extent held all things in common. There is talk about early-Christian communism. This has occasionally been counted as one of the historical proofs that communist society is possible. The corrupting of Christianity by its participation in the Roman Empire's system of domination appears from this socialistic point of view as one of many setbacks that were, according to Marx, inevitable under early economic conditions, as the demise of a premature attempt at communism.

To be sure, the members of the primitive Christian community did not live as mendicant friars, but they represented a minority, a role in which they could wholly and earnestly apply the precepts of the Sermon on the Mount to themselves. The problem of the *Catholic Church* was its historical victory, its transformation into a great, and finally a dominating religious community. The moral duty to assume political responsibility, a remote experience for the primitive community, became reality for it. The imminently expected return of Christ had not taken place. The Christians had to administer a world, whose end they were hoping for. In retrospect one might put it sharply: No one changed the world of history world as radically as the Christians, who were hoping for nothing but the end of history. This is true especially when the secularized successors to convinced Christians are acknowledged as the bearers of this Christian dynamics.[4]

[4] Cf. in this volume Chap. 5 on 'Secularization'.

This dynamics repeatedly drew its energy from the radical precepts of the Sermon on the Mount; the very fact that the church transmitted such texts, even when it was not living according to them, did not allow for any lasting calm. At the same time, though, the church had to develop an ethic, a moral theology that allowed it to give all its members rules of action to abide by—which is a precondition for a stable social structure. Some subtle nuances found in the canonized texts—such as concerning marriage in Matth. 19, 12: "Let anyone accept this who can" and in Paul (*1 Cor. 7*)—served as the basis for taking into account the insight that some people can accept for themselves stricter precepts which remain unfulfillable for the majority. This insight is constitutive in Hindu society: Each caste has different obligations to fulfill and homeless itinerant monks have yet others. The church stabilized priesthood and monasticism. The principles of monasticism were considered as counsels of the Gospel and set apart from the more generally valid precepts. The fact that a civilization with this moral and spiritual division of labor can blossom is proven especially by the Middle Ages, during which priests and monks were at the same time the most important bearers of cultural consciousness. One may point out here again: The contribution that the contemplative person makes to society consists precisely in his or her contemplation. A structure that is as questionable ethically, as unenlightened intellectually, and as thoroughly ambivalent as is true up to the present day for human society in the advanced civilizations can only prevent a decline into self-destruction, if there are always a few living within it who radically refuse to participate in its activities. And it was almost exclusively where Christians dared to do this that the spontaneous comprehension of the Sermon on the Mount flowered, such as with Francis of Assisi.

Protestants have more trouble with the Sermon on the Mount than Catholics. Protestantism is on its way to modern democratization and, owing to the abolition of monasticism, to the abrogation of the doctrine of the counsels of the Gospel, the thesis of universal priesthood, the renewed seriousness about all the commandments and—among the Reformed, at least—the presbyteral constitution. Under these preconditions, in order for the Sermon on the Mount to be made bearable, it has to be reinterpreted more clearly as a claim on one's attitude and inner disposition. A perceptive psychological observer[5] has described the compensatory distribution of roles between theology and practice in the following terms: Protestants proclaim righteousness by faith and are indefatigably righteous in their good works; Catholics trust in righteousness by good works and in this way live their faith. Such compensations at cross purposes between quarreling groups exist elsewhere too. The Sermon on the Mount reemerges as the ultimate reason, often unconsciously, behind the Protestant tension. The self-criticism, which defeated Luther in the monastery, originates from a level of insight taught in the Sermon on the Mount. The solace, the salvation that he found, obviously did not come from the words of Jesus, but from those of a successor, who himself had already encountered a problem similar to Luther's own: from St. Paul with his teaching about righteousness through faith.

---

[5] W. Kütemeyer, orally to the author.

The Christian tradition regards moral problems as a question of the personal relationship between God and man and hence under the rubric of righteousness.[6] As with all great symbols, this personal interpretation expresses phenomena of central importance. However, the continued rational interpretation of the symbols, which makes up such a great part of historical theology, gets lost in irresolvable problems because symbol and rationally comprehensible reality get inextricably mixed up and the truest, most relevant theological doctrines must then take on the form of a paradox. Thus, it became obvious to Luther that monastic asceticism cannot be justified as an accomplishment that a human person offers to satisfy the will of God. Its real purpose, namely to facilitate the attainment of an inner state that throughout history has been accessible to but few people, was thus repudiated, along with righteousness through good works. Righteousness through faith does not, however, rescind the law; and consequently the ethics of the Sermon on the Mount—Jesus' interpretation of the law—de facto became a precept for all believers. This lead on the one hand to the particular dedication to vocational work among Protestants-and thus to righteousness through good works—including their contribution to the development of capitalism, as Max Weber has described; and on the other hand to the comfort for the human conscience derived from religion as inwardness.

Sects come in multiple forms, but one of their frequent motives is, that they, under the immediate impact of the words of the Sermon on the Mount, began to put it into practice. By doing so, they almost automatically resumed the minority role that the text originated from. They created for themselves the situation of persecution, which suits the text's rules of conduct. They could thus discover that this conduct, whenever we intelligently trust the guidance of the spirit, really does lead to peace, as is demonstrated for example by the history of the Quakers. The inner transformation of the persecutors, which is the true miracle of mission, obviously lies merely as a potentiality in the drama of the persecutors' inner psyche and therefore, from the point of view of the persecuted, not in his own hands, but God's.

Modernism within Christian churches is an internal reaction of the churches to the irresistible overpowering of traditional Christianity by the enlightenment. In most cases it is as much a defensive position in retreat as is true for orthodoxy. Its specific task could perhaps be described as follows: The rationality of the modern world of will and reason cannot come to terms with its self-produced problems. The rationality of our traditional religion nevertheless remains subordinate to it, because it has not been able to distinguish between symbol and reality in correspondence with the contemporary level of consciousness. Modernism would have to inquire undeviatingly into the reality content of symbols which is accessible to modern rationality; in doing so, it would transform present-day rationality itself; it would force it to leave all its prejudices behind. This present book[7] also ventures to

---

[6] Cf. "Levels of Christian Theology: A Comment on Luther", in this volume Chap. 9.

[7] Reference is made here to the book "*Der Garten des Menschlichen*" (Hanser Verlag, München 1977); English edition (in parts), in: "*The Ambivalence of Progress*" (Paragon House, New York 1988).

take one step along this route. Modernism is, in fact, as multifarious as modernity. Liberal modernism is mostly far removed from the Sermon on the Mount; socialist modernism, which we have encountered during the last decade particularly in churches of the Third World, faces the question of whether socialism has a proper understanding of the Gospel.

Thus we are back at our starting point. The deep ambivalences of *socialism* have been one of the topics in the first chapter of this book. Socialism, which is perhaps our destiny, will necessarily become barbarism if it continues to believe that the just order can be created by will and reason alone. The quest of the younger generation today for *Asian religions* follows the right instinct; they are seeking exactly that substance lacking in the Enlightenment. But the simple classification of Jesus under the category of incarnations makes use again of symbolism, where reality would be necessary. The closer look reveals that Asian religions have already been changed by their encounter with Christianity and the enlightenment. No Hindu of former times would have talked like Vivekananda or acted like Gandhi. In religion, too, the path leads into a future for which the past cannot serve as an example. What Jesus had called the kingdom of the heavens is still possible today and is not yet completed. Two thousand years is not a long time.

Prof. Dr. Carl Friedrich von Weizsäcker with his wife in Japan (1974). *Source* The photo collection of the C.F. von Weizsäcker Society represented by Dr. Bruno Redeker who granted permission to use this photo. Original source unkonwn

Chapter 7
Thoughts of a Non-theologian on the Theological Development of Dietrich Bonhoeffer

Dietrich Bonhoeffer was a political martyr. Through his life and word he revealed the Christian faith in resistance to German National Socialism. In World War II, he joined the circle of secret resistance among the German elite that eventually attempted the assassination of Hitler on July 20, 1944. He was executed, along with many others of this group, by National Socialist justice.[1]

Bonhoeffer was a highly gifted, conscientious, intellectually clear, and unwaveringly progressing theologian. In the decades since his death, the stations of his theological development have had a tremendous influence on German-language and ecumenical theology. Where would he stand now, if he lived? In any case, we study his theology because it is relevant to our own problems.

These are thoughts of a non-theologian about Bonhoeffer's theological development. The word non-theologian is to designate here a member of the church who had no formal theological training and holds no theological office. It designates especially someone whose intellectual work takes place primarily in fields that traditional Christian theology considers to lie outside of theology: science, philosophy, and long-range politics. On the basis of my knowledge, I cannot and will not avoid seeing Christian theology and the church itself, which is my home, in its total cultural conditionedness. I believe that this perspective belongs to what Bonhoeffer envisaged in his last period as a possible development under the somewhat ambiguous label of 'religion-less Christianity'. I come from the same social milieu as Dietrich Bonhoeffer, and am only 6 years younger than he; but unfortunately, I never met him in person. I have met many of his followers and friends, mostly after the war. My main source for this lecture is, apart from Bonhoeffer's own writings, the comprehensive biography by Eberhard Bethge.[2]

---

[1] This text reproduces a lecture held at a symposium at Geneva in February 1976 commemorating the 70th birthday of Dietrich Bonhoeffer. It was first published in English in: C.F. von Weizsäcker: *The Ambivalence of Progress*. (New York: Paragon House, 1988): 263–278; the German original is part of C.F. von Weizsäcker, *Der Garten des Menschlichen* (München: Hanser Verlag, 1977): 454–478. All rights of the Weizsäcker Bequest were transferred to the Carl Friedrich von Weizsäcker Society, a nonprofit society whose executive chairman, Dr. Bruno Redeker, granted permission on 5 March 2014 to use selected texts of Carl Friedrich von Weizsäcker for this volume.

[2] Eberhard Bethge, *Dietrich Bonhoeffer: A Biography*. London: Collins 1970.

K. Raiser (ed.), *Carl Friedrich von Weizsäcker: Major Texts on Religion*, SpringerBriefs on Pioneers in Science and Practice 24,
DOI: 10.1007/978-3-319-03704-2_7, © The Author(s) 2014

7.1 Influences on Bonhoeffer's Theological Development

Bonhoeffer's presuppositions, the influences that formed him, can be loosely grouped under four headings: his family background, Harnack, Barth, and ecumenism.

Dietrich Bonhoeffer was one of those *homines religiosi* whose decision to place their own lives in the service of God comes early in childhood, unnoticed by others. There were some theological ancestors in his liberal, scholarly, upper-middle-class family; the well-ordered but easy-going family still included the children's evening prayers, even if they did not go to church on Sunday. To the circle of eight brothers and sisters, the firm decision of the youngest brother to become a minister seemed rather unusual and not entirely understandable. A child can have a silent and intense life with God, without doing violence to his natural child development, where the environment offers him only the cultural forms in which he interprets and unfolds his inner experience. It seems to me that the source of Dietrich Bonhoeffer's life to the end of his existence was such an experience opened in childhood and perhaps never fully revealed to others.

His family gave him something different that was important for his later intellectual development: modern consciousness in the form of tradition. Modern consciousness: The dominant member of the family was, until the end, the father, who was a psychiatrist at the University of Berlin, a scientifically trained physician. Unmistakable, in his written remarks, was the wise, cautious, self-critical, and yet sure style of this man. There was also the oldest brother, Karl-Friedrich, the leading physico-chemist of his generation in Germany, and there was the world of liberal-constitutional politics of his brother Klaus and of his brothers-in-law Schleicher, Dohnanyi, and Leibholz. Many theologians of our century have never fully assimilated the modern consciousness. Others had to acquire it as a rupture in their tradition with the resulting pain, tension, and doubt. Still others, perhaps, never even knew any kind of spiritual tradition in their childhood but grew up in the reduced dimension of the self-satisfaction of secularism, encountering our tradition only later, almost like a foreign culture. Dietrich Bonhoeffer never had to experience a break with his family, and it was in this continuity that modern thought was natural to him.

The choice of the religious way of life, for the son of a bourgeois, academic, Protestant family, meant the study of theology. The first of the theological influences that affected him was Harnack. Adolf von Harnack was a Berlin colleague of Bonhoeffer's father, and a Grunewald neighbor of the family. He was the most famous and most important of Bonhoeffer's personal theology teachers. His name may be understood here as representing the world of scholarly theology in the sense of historical and philological research. This research into biblical textual criticism, this critical church history, and this putting the history of church traditions in the context of profane historical description were the result of the application of modern consciousness to theology. The dogmatic position of this school is known in the theological tradition as 'liberal.' In fact, the rational analysis of the past liberated thought from the norms of the positive, from the burden of having to believe what was said merely because it had been said. Harnack's history of dogma taught that the classical Christian dogmas must be understood as formal means of resolving debates that resulted from early Christian tradition, but which could only be understood in the language of Greek philosophy as the conceptual framework of the culture of late antiquity.

Bonhoeffer, with his great receptivity, absorbed the rationality of this theology as a self-evident body of knowledge. But his own scholarly interest in theology was not historico-philological, but rather, what protestant theology calls systematic. The philosophical concept of 'system' that resonates in this word does not really express the actual concern, especially Bonhoeffer's. Simply stated, what mattered to him was not who did or said what, but rather what was true. This sought-after truth was for him, from the beginning, never the form of a statement, but that of real life. I see him as someone who, though a conscientious student of what was expected, made from the beginning his independent quest for reality. The liberating impetus at the beginning of his journey was the theological awakening of that decade that is best represented by the name of Karl Barth, notwithstanding a polyphony of other names.

The strength of liberal theology lay in the exposure of the profanely knowable in Christian history. In its questioning about religious reality it walked intellectually near an abyss. If rational thought leaves the classical self-interpretation of the Christian belief behind, does a better interpretation of this belief remain, or does belief disappear along with its old forms? If you were enlightened and remained a theologian, you had to hope for the better interpretation. Where this interpretation was not intellectually flat and thus ephemeral, it could not avoid conceptual borrowings from philosophy, no longer from the original Greek forms, but rather the modern. What seems to have survived amidst the multiplicity of opinions has proven to be the recourse to the subjective, to the manner in which religion is to be experienced, from Schleiermacher's feeling of absolute dependency to Bultmann's reservation of the existential for faith confronting nature, the latter being easily devoured by the objective sciences. This theology furnished to the church in the bourgeois period a self-interpretation corresponding to its social role. The church was seen as a necessary member of a *de facto* pluralistic society, as an essential and yet relatively non-committal guarantee of meaning; the theology of subjectivity was flexible enough to make its demands on the inner life broader or narrower, without having to perceive itself as dishonest. The liberal state had problems only with the Catholics and a few Calvinist fundamentalists who did not have a theology of feeling.

According to Barth's own admission,[3] it was a political shock that freed him from Schleiermacher. As he was a leftist Swiss bourgeois, his eyes were opened, from the beginning, to the moral intolerability of the public identification of the leading German theologians with the pompous nationalist slogans with which the threatened Germany of Kaiser Wilhelm concealed its fear at the beginning of World War I. He read again the Reformers and the Epistle to the Romans, which for Luther had been the decisive text in the New Testament canon. He discovered for himself the 'vertically descending' word, the word of God spoken through and in Jesus Christ to our subjectivity, that is, a revelation that cuts through all our general philosophical concepts, brought by a unique human being. After the war, the Protestant churches of continental Europe were receptive for this seed. How could Bonhoeffer not have been influenced by it? How did he react to it?

The fourth influence on Bonhoeffer, I said, was ecumenism, in the original sense of the word referring to the entire inhabited world. As were many others, Bonhoeffer was liberated through the ecumenical experience from a provinciality that had been the result, for almost all churches, of schisms, of nationalism, and especially of the factual experience that the thinking of the church was becoming irrelevant as a consequence of the enlightenment. Or perhaps he sought the ecumenical experience earlier than most of his religious contemporaries because, as a modern man, he had been stifled by provinciality from the beginning. He sought not only the universal reality of the church and the multiplicity of serious beliefs in the churches of the world united by the ecumenical movement. He was interested in the profane and social reality of the countries where he stayed. He went to Rome to learn about the Catholic Church at its center. What would have happened to him if he had been able to make his long-dreamed-of journey to Asia, if he had been able to speak with Gandhi, if he had been able to experience the reality of Hinduism and Buddhism, with his perceptive organs?

7.2 Christian Commitment and Modern Consciousness

I must proceed a bit further in my thoughts on the formative influences. In all four aspects, family background, Harnack, Barth, and ecumenism, what was at work was the contrast between Christian commitment and modern consciousness. The biographical and historical situation in these two realms was such that the Christian commitment was the given and the modern consciousness appeared as the environment, or influence, or partner, in the dialogue. As a non-theologian I would like to make a methodical inquiry and reverse the direction of the questions. One cannot understand the modern world or the development of Christian theology of the last decades, and Bonhoeffer's influence, if one is not prepared to accept modern consciousness as the given, and the Christian commitment only as an

[3] Cf his postscript to '*Schleiermacher-Auswahl*', Hamburg/München 1968, p. 293.

element of an almost forgotten earlier history and possibly an initially confusing and perhaps disturbing partner in the dialogue.

It seems quite obvious to me that the modern consciousness of Dietrich Bonhoeffer came from the fact that his family life included emphasis on science, even if neither he nor his later surroundings had particularly scientific interests. Science is the hard core of modern thought. But consciousness of science has not been the leading factor in modern intellectual history, as this may have been the case for certain culturally limited phases of the 17th, 19th, and 20th century. Yet science is the most resistant intellectual product of modern times, its hard core. One could say in an inquiry into historical materialism that the laws of science are the objective background of the economic and the economy-dependent cultural progress of modern times, even where this fundament remained concealed or seemed unimportant to the subjective consciousness of the bearers of this progress. Where would the dominating and artistic subjectivity of the renaissance person have been without the urban culture made possible through the late-medieval technology? Where would the idealism and romanticism, as well as the deliberate realism, of the 19th century have been without the bourgeois way of life made possible through the technology of the industrial revolution? And on the theoretical level, the father of strictly subjectivist philosophy, Descartes, reflected upon the *res cogitans* in order to comprehend how he could think about nature as *res extensa,* that is, with the help of the concepts of mathematical science. Those who, like Dietrich Bonhoeffer, have lived with scientists know something of the imperturbability of modern intellectual progress.

The historico-philological theology was part of the second great scientific advancement of modern times, the discovery of a historical way of thinking or, in a more general sense, of the humanities. The accomplishment here is the ability to understand alien individuals, even alien cultures, by *not* identifying with them. This is an enormous expansion of our intellectual horizons. It was actually only since Wellhausen that the Old Testament was rediscovered in its unbelievable richness and intensity and its relevance for our own decisions, that is, since Old Testament scholarship taught us to see the cultural and social conditions that prevailed in Israel 3000 years ago and to which the old texts correspond with immediate liveliness and without any Christological reinterpretation. I dare say that the confusing state of New Testament exegesis—confusing in comparison with Old Testament scholarship—will not change until one learns to accept that pivotal texts like the Sermon on the Mount were addressed to people who lived like mendicant friars. Anyone who is familiar with the teachings of Buddha or remembers how directly St. Francis of Assisi was able to read these texts knows what I am talking about.

All three theological examples that I have referred to above should oblige us to cross the threshold that separates the peaceful science of Harnack from the unrest symbolized by the early Barth. Contemporary humanities mostly achieved their enormous expansion of horizons through a neutralization or, even worse, a trivialization of its own value judgments. The so-called cultural Protestantism, the justification of Christianity through its contribution to the prevailing cultural setting, includes some very good historical insights, and yet it is a trivialization that ignores the abyss that enlightened theology is passing along. If Christianity is justified through its cultural contribution, then every cultural criticism, every world reform, be it social-Darwinist, or technocratic, or Marxist, will let the Christian faith sink further into forgotten and remote history. Barth knew exactly that, if he was right in following Christ in his life, Christ must be more than all these world reformers and, in particular, more than the world that permits itself to be so changed. I said to Karl Barth, in the one very long conversation I had with him, at the beginning of the fifties, that I saw the direct path leading from Galileo to the atom bomb and that I was confused by the question of whether I could continue, with this knowledge, to pursue my beloved physics. He answered: "Herr von Weizsäcker, if you believe what all Christians avow and almost no Christian believes, namely that Christ comes again, then you may, you must, continue with physics; otherwise, you must not continue." That is how he had to talk. He also said to me, when we came to talk about Gogarten: "He may find my way of speaking too mythological, I find his too philosophical." I trusted in the truth in Barth's mythology and have never given up physics.

But after this agreement with Barth's position, the non-theologian is now obligated to voice all possible skepticism. It is astonishing enough that the biblical texts are still able to speak so demandingly to enlightened persons today. But they reach, almost only, people who were still educated in the language of the church, with dialectical theology remaining a subject for church intellectuals. The non-intellectual members of the church keep falling away in our century, decade by decade, at least in the old Christian nations, for the apparent reason that also the intellectual bearers of modern consciousness in these countries distance themselves more and more from Christian content. The church has not been able to overcome its provinciality in comparison with the openness of modern consciousness. It appears to me to be structurally clear that it cannot overcome this through orthodox insistence or through a modern version of apologetics. The questions of content that present themselves here have, it appears to me, determined not only Bonhoeffer's theological development but also his theological legacy.

7.3 A Journey to Reality

To the extent that I have been able to survey the secondary theological literature of the past decades about Bonhoeffer, I see that it interprets his theological development as immanently theological, which means, referring to its validity for other theologians. For example, this is true—despite their broad historical horizons—for both Bethge's biography and Ernst Feil's theological monograph.[4] I cannot, and do

[4] Ernst Feil, *The Theology of Dietrich Bonhoeffer*, Philadelphia: Fortress Press 1985.

not want to, compete with these thorough and penetrating studies. I follow Bonhoeffer's theological development only in loose reflections. I said that I find his development to be a journey to reality. It is the reality that would prove itself to be the basis of the church, beyond its provincialism.

Let us proceed now to Bonhoeffer's five books, *Sanctorum Communio, Akt und Sein* (Act and Being), *Nachfolge* (The Cost of Discipleship), *Ethik* (Ethics), and *Widerstand und Ergebung* (Letters and Papers from Prison). The first two were written for the acquisition of academic degrees and could be called philosophical works; the next two are works in ethics, and the last contains the letters written in prison, which I see as Bonhoeffer's breakthrough to his own origins. This division corresponds somewhat to Bethge's division of the phases of Bonhoeffer's life into "The Magic of Theology," "The Price of Being a Christian," and "Participant in Germany's Fate."

Bonhoeffer's first writings, which were hardly noticed at the time, anticipate the fundamental thoughts of his last period. The dissertation *Sanctorum Communio* is, in two ways, the beginning of the journey towards reality: beyond the liberal theology and beyond Karl Barth. It is a systematic, I would even say philosophical, work in the field of theology. Its contents include neither concepts of individual religious behavior such as the concept of belief, nor concepts of religious metaphysics such as the concept of God, but its theme is the concreteness of religious existence in the social framework of the church. And methodologically speaking, it does not deal with theological postulates, but proceeds from the profane science of sociology and attempts to prove the theology of the church as the actual, the only, enduring fulfillment of the socio-philosophical teaching about community. "Christ existing as community" is the formula of this fulfillment. His professorial qualification thesis, *Akt und Sein* proceeds directly from the contemporary philosophical controversy about the tension between the understanding of the self, of the human being, as a transcendental subject—expressed in the title as 'Akt' (act)—and the ontological understanding of the human being as 'Sein' (being). He sees both ways of understanding failing in their isolation and finds their possible unity, again, in the church.

The young Bonhoeffer accepts the contrast—prevailing in his time—of a phenomenological-apriorist social philosophy and empirical social research, the empiricism supposedly based on the results of the phenomenology. He identifies the *a priori quality of the* structures of the community (different from those of society) with the Christian-theological concept of the original condition, in contrast to which exists the real historical community, in the shadow of the destructive force of sin. The church, Christ as community, is historically the only possible way to live as a community under the aspect of sin. Today hardly any social philosopher would dare place a phenomenologically provable a priori before the social experience. The historical conditioning of all social forms is only too well known to us, and thus the anthropology that lies behind this history has become an irresolvable problem. The equating of philosophically ideal types with the original state of humans before the fall as told in the Bible may be an effective thought in the history of Christian reflection. But one can make it understandable only historically, through the two sources of Christian theology: Greek philosophy and the Bible. The word 'theology' itself comes from the religion of Greek philosophy and indicates the core of this philosophy. The opposition between theology and religion as two separate spheres of knowledge results from a later period of Christian thought, medieval Scholasticism. It is an attempt to determine the reciprocal relationship of the two religions, the biblical and the philosophical, that converged in the intellectual history of Christianity—an attempt whose conceptual apparatus came from one of the two, namely, from philosophy. The systematic distinction between theology and philosophy exists, to the best of my knowledge, in no world culture except that of the second Christian millennium, that is, the medieval-modern European millennium, which is now coming to an end.

In my subjective opinion, the young Bonhoeffer is not to blame for not seeing the historical relativity of his problem, and perhaps even consciously repressing it. He had to defend himself against another temptation of contemporary Protestant theology, in which the influences of Harnack and Barth work together, namely, the tendency to eliminate Greek philosophy and thus, *de facto,* philosophy as such out of the self-understanding of Christian belief. What would remain, if this attempt were successful, could no longer be labeled theology. This opinion rests on my conviction of the nontrivial continuity of philosophy.

If Bonhoeffer had remained an academic theologian, it seems to me that he would not have been able to resolve the problems he dealt with. In a free decision and in reaction to historical necessity, he chose a way that is more real. As a result of Hitler's seizure of power, the Protestant church in Germany came out of its condition of irrelevance and entered one of distress. Bonhoeffer proved that he meant what he said when he made the church the central concept of his theology. He soon became one of the most decisive protagonists of professing Christians of radical orientation. This cost him his academic career and brought him to the experience of the living community, first with the young theologians in the minister's seminary in Finkenwalde and later with its successors in Eastern Pommerania.

Bonhoeffer's life from now on did not proceed, as tradition and talent would have had it, into the wide world, but rather into increasingly narrower confines, from the wide Atlantic world back to Germany, from the university to a minister's seminary, from there to the temporary hospitality of the monastery of Ettal and, after a brief excursion into the highly secretive community of political conspiracy, to prison, and, after a waiting period of 2 years, to his death. In the first eight of these 12 years, one has also seen, not without reason, a certain theological contraction. The professing church, and with it its supporters, shows a certain intransigence. And the theology of 'The Cost of Discipleship' speaks a biblicist language. I believe, though, that Bonhoeffer in both respects followed his inner voice towards reality.

In 1933 I was too removed from church organizations and too apolitical in my actual awareness to perceive what I discovered, with a kind of horror, 40 years later in Bethge's book: how the majority of the leading clergy in Germany initially were taken in by Hitler's national-conservative rhetoric, or how they at least thought it necessary to pretend that they were. If at least they had been impressed by

Hitler's revolutionary tone! Actually they were, as were most of the conservative bourgeoisie, useful idiots in support of Hitler's real intentions. Bonhoeffer, thanks to the political horizons of his earlier family life and his sophistication, was compelled to assume an opposing position. The political horizons of many of his colleagues in the clergy, it is true, were no wider than those of his opponents. With his fixation on confrontational struggle Bonhoeffer probably was unable to appreciate the wise, spiritually and not politically grounded ability of generous comprehension in the rather exceptional person of the elected presiding bishop Fritz von Bodelschwingh, who had the ability of a loving relationship with emotionally disturbed patients, also in politics.

The biblical theology of "The Cost of Discipleship" has been given shape in the context of the National Socialist experience, but it was not triggered by it. If the church was "Christ existing as community," then it had to ask, "What, after all, does Jesus want from us?" Bonhoeffer, in distinction to many theologians, did not hesitate to interpret the central text of the Gospels in a direct understanding. He was educated and intelligent, but more important for him was to see and speak about reality so simply that it becomes binding. Thus his breakthrough to the Bible did not occur in the encounter with the highly intellectual Epistle to the Romans, but in the unbearably merciful simplicity of the Sermon on the Mount.

If we ask what Bonhoeffer has understood of the abyss that this text conceals, we see that his exegesis is not consciously learned but shaped almost like sermons. The place for these sermons is first the preaching seminary, whose participants prepare themselves for service in a church that is persecuted or close to persecution. This framework is close enough to the condition of Jesus' disciples to overcome the almost insurmountable hesitation in an established church about a literal interpretation of this text and to make a spontaneous understanding possible. Bonhoeffer had not hesitated to urge the community of his candidates for the ministry, who were almost of his age, to assume certain ancient, helpfully tested rules of monastic life, such as a regular daily routine, a liturgy of prayer, and even some modest exploration of the rich experience of meditation. If he could accomplish this in opposition to deeply entrenched Protestant preconceptions, then I see this as exactly the same courage in approaching reality that we find in his later theology of the secular. His fate should demonstrate that the Christian life can only become more worldly if it becomes more spiritual and only more spiritual if it becomes more worldly. Both of these steps amount to the elimination of the protection against one's fear of one's self.

Bonhoeffer's courage to relate every word of the text as a liberating commandment to his own life has unlocked this text to him, and to all who read him, far beyond the customary reading. In this he encountered an unavoidable ethical problem, actually a problem of becoming conscious that everyone faces when reading the Sermon on the Mount with the sincere will of imitation, without being able to assume the living conditions of mendicant friars—whether this inability results from not wanting or from being prevented to assume these conditions. It appears to me that Bonhoeffer, in this period, did not have the right access to an entire layer of this text, nor—as far as we can tell from his written testimony—did he ever gain it later. In terms of positive ethical consequences, the insights of "The Cost of Discipleship" point to pacifism, a position which he gradually adopted under the influence of his friend Lasserre. But this orientation of political thought separated him increasingly from the political thinking of his church, even from close associates. The social reality of the Protestant church in the bourgeois era permitted almost only the choice between agreement and avoidance with regard to official politics, two politically unproductive modes of behavior. What I have called Bonhoeffer's awareness of the condition of the world led him from these narrowing horizons into active political participation. If we compare passages of his 'Cost of Discipleship' with the conspiratorial political activity of his last years in regard to the questions of truth and secular objectives, we may not be able to reconcile them fully. He has personally accepted the ethical conflict between the positive ethics of confessing discipleship and the positive ethics of political conspiracy, and finally decided himself in favor of politics. After that, he never wavered. His political action and his suffering in prison, which resulted from it, opened without doubt a new dimension of reality to him.

The first phase of the development of this consciousness of reality is reflected in the fragments of his *Ethics*. The difficulty is not the insight into political reality, which was always clear to him; what is difficult is not to lose the Christian reality through immersion in political action, but to understand it in an even deeper sense. Without this experience he would probably not have written in [at the beginning of] his *Ethics*[5]: "The knowledge of good and evil appears to be the goal of all ethical reflection. The first task for Christian ethics is to supersede that knowledge... For Christian ethics, the mere possibility of knowing about good and evil is falling away from the origin [Ursprung]. Living in the origin, human beings know nothing but God alone. They know other human beings, things, themselves only in the unity of their knowledge of God; they know everything only in God and God in all things. Knowledge about good and evil points to the prior disunion and estrangement [Entzweiung] from this origin" (p. 299f).

This rediscovered unity speaks in every transmitted word of Jesus (p. 312); Jesus refuses to be drawn into the conflicts of life (p. 313); "do not judge so that you may not be judged" (*Matt. 7.1*), because judging itself is apostasy (p. 316). Now Bonhoeffer's ethics turns from the last things to the "next-to-last things" (p. 152). "Christian radicalism, whether it would flee the world or improve it, comes from the hatred of creation. The radical cannot forgive God for having created what is" (p. 155). "Thus, the concept of the natural must be recovered from the gospel itself" (p. 173). Bonhoeffer's ethics turns to concrete details of life and, in history, to the real sequence of eras ("Heritage and Decay," p. 103ff). "The world remains the world *because* it is the world that in Christ is loved, judged, and reconciled.

[5] Weizsäcker here quotes from an earlier version of the German text in which this section was placed at the beginning of the book. Meanwhile a new, critical edition of Bonhoeffer's *Ethics* has been published as Vol. 6 of Dietrich Bonhoeffer's Works (1992); an English translation of this new edition has been published by Fortress Press: Minneapolis in 2005. Page references are based on this new edition.

No one is commissioned to leap over the world and turn it into the kingdom of God" (p. 267). At that time, Bonhoeffer regarded *Ethics* as his probably final, fulfilling theological work. That it remained a fragment is certainly not only a consequence of the political circumstances of his life. The open cracks of the fragment point to the missing step, the one he took in prison.

The breakthrough, as we can see in his letters to Bethge, occurs after his first year in prison; it happened with the force of a landslide. Judging by style differences, I would say that the last scales had fallen from his eyes; he now sees the reality that he had always felt. "We are approaching a completely religionless age; people, as they are now, simply cannot be religious anymore" (April 30, 1944).[6] "I often wonder why my 'Christian instinct' frequently draws me more toward the nonreligious people than toward the religious, and I am sure it's not with missionary intent; instead, I'd almost call it a 'brotherly instinct'... When others begin talking in religious terminology, I become almost totally silent" (April 30). "I consider the attack of Christian apologetics on the world's coming of age as, first of all, pointless, second, ignoble, and, third, un-Christian" (June 8). "Barth was the first to recognize the error in all these attempts... to save some room for religion in the world or over against the world,...but in the nonreligious interpretation of theological concepts he gave no concrete guidance..." (July 8). "I will write to you in Italy about the Song of Solomon. I would in fact read it as a song about earthly love, and that is probably the best 'Christological' interpretation" (June 2). Bultmann's essay on the "demythologization of the New Testament" did "not go 'too far,' as most people think, but rather not far enough" (May 5). "God is the beyond in the midst of our lives" (April 30). About W. F. Otto's book on *Die Götter Griechenlands* (The Gods of Greece): "Can you understand that... I am—*horribile dictu*—less offended by gods who are so portrayed than by certain forms of Christianity? that I almost believe I could claim these gods for Christ?" (June 21). I am touched to see that from my book *Zum Weltbild der Physik* (The World View of Physics) he quoted the thought "that we shouldn't think of God as a stopgap for the incompleteness of our knowledge" (May 29), and that he added: "if only a spiritual exchange were possible" (May 24).

Bonhoeffer saw. The manuscript that he began on these thoughts is lost, but I doubt he would have been able to work out the interrelated concepts of "religionless Christianity," "non-religious interpretation," and "arcane discipline" (May 5) in such a way that his thoughts would have done justice to his vision. Bonhoeffer's final phase is important to the church as a destruction of a defense position, not as the basis of a theology. In a mysterious way this phase has the characteristics of a work of mature age: a widening of perception in the proximity of death, which indicates in words the steps that the speaker has taken forebodingly, although in the world in which he has lived one is not able to take these steps.

---

[6] References to dates and the English translation of the quotes follow the new, critical edition of the "Letters and Papers from the Prison", published as Vol. 8 of Dietrich Bonhoeffer's Works, Fortress Press: Minneapolis 2010.

7.4 Religion and Modern Consciousness

What does religion mean to us?

First: How does religion face modern consciousness? Before we go on, let us reverse the critically restrictive use of the word 'religion' with Barth and Bonhoeffer. Let us open our eyes to the wealth of phenomena that can be subsumed under the general concept of religion. Without any systematic pretension I would try to group these phenomena under four aspects, i.e. to distinguish four dimensions in religion: religion as element of a culture; religion as foundation of a radical ethics; religion as inner experience; religion as theology. We have encountered examples of all four dimensions in this lecture.

Religion as element of a culture, even as the bearer of a culture, this is what Christianity has been during these past 2000 years. Religion as bearer of a culture shapes the social life, structures the times, determines or justifies morality, interprets anxieties, shapes joys, consoles the helpless, gives meaning to the world. It is the traditional form of this religion that is slipping from modern consciousness.

Religion as foundation of a radical ethics is critical of, and therefore often persecuted in its culture. Wherever the Sermon on the Mount or at least the Ten Commandments have been taken literally, religion was the basis of a radical ethics.

Religion as inner experience is, in a way, any subjective experience of the two foregoing aspects of religion. It is living consciously in faith. But it is, especially, prayer, meditation, mysticism.

Religion as theology is the attempt at understanding the experience of the other three aspects in a conceptual way. The thoughts we are sharing in this lecture are a part of theology.

If we want to define a general concept that comprises these different phenomena, then we could say: Religion is human conduct in encounter with the divine. This definition seems to sacrifice the neutrality with which we described the four aspects. In fact: If we include in a definition the concept of the divine, how, then, do we define the divine? Formally speaking one could answer: This is the philosophical circle from which no fundamental reflection can escape. A definition of truth claims to be true, a definition of philosophy is itself philosophy, a definition of religion is theology as the attempt to understand the experience that we call religious. The formal answer, however, is to lead us on to a more fruitful substantive answer. This precisely is the amazing and reflection-provoking phenomenon: that the entire human history is overshadowed by a mode of experiencing and thinking that can interpret itself only as human conduct encountering a divine reality.

Let us now turn to the various manifestations of religious criticism. There is a philosophical critique of religion that today often considers itself as scientific. The rationality of traditional religion is no longer convincing. If one were to ask what is the basis of the possibility of religion, one would either be led to conclude that, in view of the arguments of rationalistic metaphysics, the historical form of religion, its role as bearer of a culture, and the experiences of religion are mere wrappings, or to reduce all religious experience—through radical skepticism—to the status of self-deception or deceit.

The political pathos of religious criticism does not originate in theoretical philosophy, but in morality, in radical ethics that is independent from religion. *Tantum religio potuit suadere malorum,* says Lucretius in *De rerum natura,*[7] about the many evil deeds that religion can make us do. The classical attack by the Enlightenment against religion aims at its role as element of a culture, that is, as bearer of the relationships of power and domination. It is not unbelief, but rather morality that is the determined enemy of religion. Anyone, who does not take this seriously, does not understand the history of the Enlightenment. This opposition exists also, in latent form, in the compromises—as represented by the rationalistic metaphysics of the 17th and the 18th century—that established, after the religious wars, a natural law that remained neutral with regard to confessional positions. The compromise disintegrates in the course of history, and the Marxist phase of the Enlightenment considers it to be the dominating ideology of the bourgeois era.

The criticism of religion in the old Asiatic cultures is characteristically related and yet different. Confucianism does not see itself as a religion, but as a rational morality that bears a culture; and this is the reason why it appealed so much to the representatives of the European enlightenment of the 18th century. But Confucianism, too, is a conceptual compromise between morality and the religion of heaven and stabilizes domination in a system of deference; Mao's critique of Confucius radicalizes morality as egalitarianism.

The great meditative teachings of Hinduism and Buddhism consider their own religion in its exterior manifestation, which bears the culture in which they live, merely a representation of the truth proper through the anthropomorphic, psychically conditioned imagination of deities. Religion is recognized here as an element of culture, but on a lower level. On this path, radical ethics, often heightened to asceticism, is required, though not as a law as such, but as an ineluctable stage of purification in order to become receptive for the decisive experiences. The theology of these schools is an interpretation of what is experienced in meditation.

Where is our own position between religion and morality? To contemporary consciousness this is an anthropological question, a question of understanding the human condition. If we are to gain understanding, then we must not play down the contrast between religion and morality but consider sharply its character as an ideal type. Enlightenment, understood as self-enlightenment, as liberation from self imposed immaturity, is a respectable undertaking. The theoretical pathos of the Enlightenment meets with morality in the virtue of truthfulness, whose core consists in not lying to oneself. Lucretius blames religion for *suadere,* for persuading; truthfulness makes Bonhoeffer become silent when others begin to

[7] Lucretius Carus, *De rerum natura, I, 102.*

talk in religious terminology. When the autonomous ethics wants to be truthful, it becomes radical. This is not a matter of an abstract consequence; a realistic look at human society is sufficient. The problems of justice, of treating others humanly, even of survival remain unsolved as long as each individual, and even more each social group, applies the ethical demands strictly only to others, but to one self rather laxly. The result is the chain of catastrophes that are called political history. Our human behavior in this history is not basically evil; it is, above all, beneath our own intellectual level, it is stupid. There is hardly any human conflict of interests that could not in principle be resolved on the basis of reason. The stupidity that prevents a solution stems from a lack of good will, and this lack is a consequence of fear. Any morally sensitive individual, who notices this in himself, has to hate himself. But every failure of our alertness, of our truthfulness, will allow us to project this hatred onto others. For this reason, the struggle for justice is accompanied by the pathos of hate. In this way, however, those struggling for justice legitimate the fear and hatred of their adversary. This affective entanglement between humans becomes stabilized through mutual fear. It can be overcome not through rational analysis but only through a purer affect, through love. The experience that love is possible is the religious basis of radical ethics. The possibility of a reconciling love is experienced as grace. Opening oneself toward this grace is faith. The traditional religious expression of this experience is that we can truly love our neighbor only in God. The commandment "Love your neighbor as yourself" cannot be fulfilled in autonomous morality, because I cannot conscientiously love myself but in God; the more sensitive I am, the less I should forgive myself for being the way I am.

This kind of reflection can take us to the door of religious experience, but it does not unfold this experience. People of Western culture have encountered this experience in relation to Jesus, in commitment to him, in the attempt of following him in discipleship. This experience is bliss and the sword at the same time. It opens our eyes for the way in which religion is being used in support of the framework of interests in the prevailing culture. The critique of religion on the basis of a radical ethics does not become false just because radical ethics is possible only on a religious basis.

From this insight stems the need to apply different names to the different aspects of religion. Barth chose the name 'religion' for that element of religion that is the subject of religious criticism; for the rest he chose words like 'Christianity' or 'faith.' He knew however, that he was taking on an immense task, precisely as a theologian. How could faith express itself without right away becoming again 'religious' through the concepts of language? In my conversation with Barth I remarked, in a questioning mode, that the Gospels were a rather short text while his dogmatics was long. He replied, with a smile: "Indeed, if we were as simple as the Gospels, then we could also be as brief in talking about it." He added that in his dogmatics he had only tried to work through the large amount of misinterpretations of the Gospels. However, I believe that, for its resolution, this problem requires other means than those that Barth was able to employ. Why did he choose the term 'religion' for all that was questionable in Christianity? The reason may have been that for scholars of the academic study of religion in the 19th century 'religion' was the general concept that included 'Christianity' and 'faith' as particular concepts. Instinctively he realized that the conceptual pyramid derived from logics was covering up the phenomena. Indeed, whoever has tried to communicate an experience to theoreticians will know about the certainty with which theoreticians detract attention from the experience and soften an ethical demand by classifying it in their conceptual framework. Barth wanted to repel the influx of the religion of Greek philosophy into Christianity. But one must say that his theology, too, could not do without the use of concepts, and that he did not accomplish a clarification of the relationship of theological concepts with concepts in general, thus a philosophical reflection on the kind of philosophy that was implicated in his way of thinking.

Bonhoeffer's last statements, too, fell short of the true, substantive problem. In his writings it might appear as though the 'coming of age' of the consciousness of the modern world were a way of thinking that has internal consistency. If this were true, then I could hardly imagine an interpretation of the old faith within the concepts of this way of thinking. Bonhoeffer's hints at an "arcane discipline" "through which the *mysteries* of the Christian faith are sheltered against desecration" (May 5) point in another direction, as does the claim that "the coming of age of the world… is now actually better understood than it understands itself, namely, from the gospel and from Christ" (June 8). This is also the line with Gogarten's later theology, which understands secularization—in distinction from the phenomenon that he called secularism—precisely not as decline of faith but as realization of God's will in history. All these positions should be asked how they manage to move within the circle of understanding Christianity in terms of the modern consciousness, while understanding modern consciousness from the perspective of Christianity.

The designation of the modern world as 'having come of age' is misleading when we go beyond the parallel to the legal emancipation from parental authority. It is true, modern culture is no longer subject to the control of the church. With regard to the development of consciousness one could, at best, speak of a special phase of maturation, in the sense of the formation of the interrelated faculties of will and intellect, whose common tool is conceptual thinking. But in the biography of a person as well as in the history of a culture, phases of maturation usually mean that certain abilities are repressed when others take their place. This applies especially to phases of emancipation, which for this reason often take a dramatic course. Emancipation is not an end in itself, not the attainment of an equilibrium; it is rather a transition, a release of contained forces. Contemporary emancipation has become ecumenical. It has stretched to all cultures of the world. However, it does not have the moral rules necessary to shape the liberated world. This is especially true if the thoughts about radical ethics that we discussed were correct, because modern emancipation aims at autonomous morality. Therefore, we must expect that the greatest, worldwide, political conflicts do not lie behind us but are yet to come. And one must fear that neither of the contending parties will eventually have the right—that at first seemed to be so clear—on its side.

In this situation the church preserves the knowledge of the only thing that can heal, the love of one's neighbor across the dividing lines, a love that is possible only through God. The church does not preserve this knowledge, if it persists in a claim reaching back historically to the time before emancipation. In that case it will create a further dividing line of its own, i.e. the division between church and world. Related to this division is also dividing line over against non-Christian religions. Here the task is not to love the enemy but to understand the brother. Most Christian theologians even of the rather modern generations like those of Barth and Bonhoeffer perpetuated the error that the other religions were just 'religion', in contrast to Christianity. This attitude was a defensive wall around the belief in the uniqueness of Christ, a defensive line that, like all such defenses, has made understandable precisely what is important in the words of Christ. The early church, by incorporating the religion of Greek philosophy, was wiser in this respect.

These are some of the landmarks along the way of theology into modern consciousness, a way that Bonhoeffer saw opening itself up before he died. Had he been allowed to live, he would have traveled with us on this way.

Chapter 8
Levels of Christian Theology: A Comment on Luther

In some of my essays in recent years I have referred to plateaus: in scientific history they are Kuhn's paradigms or Heisenberg's closed theories; in the organic evolution, the species; with regard to human existence, I have applied this designation to what is just, true, beautiful, holy.[1] Plateaus do not have to alternate historically, as the last examples show; they can, if they are different enough, coexist. However, as a consequence of an inner necessity in the historical process, they can also replace each other, just as a comprehensive theory explains the more specific one and thus makes it relative in its truth-claim, replaces it "as truth." In terms of the use of language, the loanword 'plateau' seems redundant; 'level' represents the same geographic metaphor.

A brief reading during holidays in Ranke's *Deutsche Geschichte im Zeitalter der Reformation* (German History in the Age of the Reformation) made me realize the importance, which the disputation with Eck in Leipzig (1519) had for Luther. In his negotiation with Miltitz, Luther had been honestly prepared to restrain himself on the issue of papal authority. But the factual strength of Eck's argument forced him, for the sake of truth, to proceed, beyond all earlier restraint, to consequences that he only then became fully aware of. Luther proved to Eck that the authority of the pope is not found in the Bible, but is the result of claims that emerged in history; he was referring to the Greek Church with its many church fathers and saints, which, although it does not recognize the pope, is seen by the Roman Church not as heretical but only as schismatic. Eck thus forced Luther *de facto* to an outright rejection of papal authority. Luther wanted to support his

[1] This text was written in 1976 following on the previous one on Dietrich Bonhoeffer. It was first published in English in: C.F. von Weizsäcker: *The Ambivalence of Progress*. (New York: Paragon House, 1988): 257–262; the German original is part of C.F. von Weizsäcker *Der Garten des Menschlichen*. Hanser Verlag, München 1977, 479–487. All rights of the Weizsäcker Bequest were transferred to the Carl Friedrich von Weizsäcker Society, a nonprofit society whose executive chairman, Dr. Bruno Redeker, granted permission on 5 March 2014 to use selected texts of Carl Friedrich von Weizsäcker.

K. Raiser (ed.), *Carl Friedrich von Weizsäcker: Major Texts on Religion*, SpringerBriefs on Pioneers in Science and Practice 24,
DOI: 10.1007/978-3-319-03704-2_8, © The Author(s) 2014

argument by reference to the councils. But Eck proved to Luther that the recent councils, in particular the Council of Constance with its condemnation of Hus, had asserted the papal authority. Now Luther was compelled to say that even the councils could err. This forced him to his real position: The only unshakable authority is the Bible. How difficult this step was for Luther is demonstrated by the state of shock into which he fell in his intellectual follow-up work in Wittenberg. Only now did he recognize that the papacy was a system that in a consistent manner expanded its power while sacrificing the truth. To the question of how God's providence could have permitted this perversion of his church, he saw only one consistent answer: The pope is the prophesized anti-Christ. One has seen this equation as evidence of excess in Luther's opposition. In reality, the equation expressed precisely an insight into the character of the papacy as a genuine 'level' in history, formulated with the help of one of the rare concepts for historical events that Christian doctrine about the history of salvation offered to him.

Whoever considers these incidents from the perspective of the Enlightenment will recognize immediately that Luther's position of scriptural authority is just as untenable as the authority of the pope or of the councils. In contact with enlightened contemporaries, such as in the dispute with Zwingli about the Eucharist, he encountered factually the next historical level: reason. Critics of the Bible discovered, in later centuries, as many contradictions in its text as in the history of the church. On another level, biblical criticism justified one of the Catholic theses against Luther: The text of the Bible is itself a part of history. If the Holy Spirit could dictate the infallible Scriptures, then it could just as well lead the church through infallible popes or councils; if he had not done the latter, why should we believe that he had done the former?

In this manner the 'level' of enlightened reason criticizes the 'levels' of Christian theology. But mere emancipation is never a real understanding of that from which one emancipates oneself. A degree of historical blindness is part of the psychological conditions of the act of emancipation. Maturity establishes an understanding of the ancestors of which neither they nor the rebel desiring emancipation were capable. This is the ('Heisenbergian') ideal case. In the history of religion, as generally in cultural history, Kuhn's image of paradigms, where later ones can no longer achieve something that the former ones could, is more appropriate than in physics. The 'levels' of Christian theology preserve treasures that remain invisible and unexaminable on the 'level' of enlightened reason.

An enlightened theory of evolution can, to some extent, do justice to the 'levels' of Christian theology. It can explain why in church history, too, there are 'levels', perhaps such intellectual 'levels'. The Christian community knew itself being led from the beginning by a living truth: "The Holy Spirit will lead you into all truth." This is an evolutionary process. As in all evolution, it proceeds through trial and error, through advances in different directions and the resulting struggles between these directions, e.g. victory of the triumphant direction, integrating agreements, and incurable divisions that open up separate evolutionary directions. Here 'levels' of decision-making offer themselves, just as in any other intellectual and political development. To the ideal of leadership through truth corresponds the ongoing process of reaching genuine agreement within the community. In order for this to be successful, one of two conditions must be fulfilled. Either the community must be so small that its members know one another and in their social relationships can test again and again the continuing validity of the true agreement; this is the principle of sects, now most visible among the Quakers. Or the truth must have such an obvious superpersonal power that even a widely scattered human group can be certain that this truth will assert itself always afresh by virtue of its convincing impact. A model for this is contemporary science, in which new insights so far have asserted themselves through active consent of the scientific community, with the peripheral phenomenon of the continual dying off of those who were not convinced. The Christian church has claimed a similar persuasive power of its truth, mediated through its preaching and the ever effective presence of the Holy Spirit. It would not have survived two-thousand-years without some justification of this claim (this is an argument for the papacy that Luther could only dispel by equating the papacy with the divinely tolerated anti-Christ). However, the sociologically and politically astute observer recognizes in the history of the church, more obviously than in the history of science, the supportive constructions that have led to the affirmation of the three authorities of the pope, the councils, and the Bible. The principle of authority arises wherever agreement on the basis of an immediately convincing truth is not successful. Authorities may be part of living experience or historical. Experienced authorities are either monarchical or corporate, that is, either episcopal or presbyteral. If the universal coherence of the church is to be guaranteed through experienced authority, then a central power becomes necessary because of the enormous expansion of the Christian oikoumene; this led to the establishment of the papacy. The principle of corporate authority can only rarely be practiced in an ecumenical framework; this led to the practice of councils. The most unshakable authority will be a definitive text; this led to the acceptance of the canon of the Scriptures.

It is evident that the simple recognition of these mechanisms, as they appear to enlightened reason, is enough to make all three authorities relative, even if one is prepared for the time being to accept them for pragmatic reasons and also out of respect. Thus Luther was prepared to respect the councils and even the pope, until Eck forced him to opt for the truth. It is perplexing to the contemporary reader that Luther stopped with regard to the authority of the Scriptures and refused to move to the next 'level', as he had refused to stay with the former one; he saw the whore 'reason' as the counterpart of the anti-Christ pope. Yet, such perplexity concerning a 'level' always shows that we have understood the 'levels' only formally as stabilizing means, and that the content that makes this stabilization possible has remained unnoticed.

The Enlightenment symbolizes its own 'level' through a half-truth, the belief in progress. It was an enormous historical step that placed the free search for truth and the belief in the possibility of a better future under the principle of reason in opposition to the generally ruling absoluteness of tradition. But even within the domain of rational knowledge, the mere belief in progress has something one-dimensional. It often is unable to recognize the substantive reasons for the validity of older insights that spanned epochs. Here Kuhn's scheme of scientific history is illuminating. Normal science does not question its own dominant paradigm; precisely because it applies it without question, it does not really understand why it functions. In the great scientific revolutions, science is forced to philosophize. Einstein understood Newton better than the eighteenth and nineteenth centuries had understood him; he could relativize him because—finally again—he did him justice. But no one who comes later will *fully* do justice to an earlier one. The great gift I received from my teaching of the history of philosophy was that I learned to see to what extent the entire modern philosophy, though pointing beyond Plato, has failed to fully understand Plato. But the veil that covered Greek philosophy was precisely its role as authority. Where this authority was still valid, it remained as misunderstood as a dominant paradigm, and where it was no longer valid, it was as forgotten as a paradigm of the past.

Christianity has never questioned the authority of Christ. It has made it sacrosanct through the dogma of his divine nature. But this authority, too, was a veil hiding his reality. This explains the ever-recurring conflictive character of the inner revolutions of the church. Whoever raised the veil a bit had to understand himself as one who was returning to the origins, to the real Christianity, and had to appear to the believers in existing authority as an enemy of Christ, as inciting a sacrilege. Thus it was for Luther. The principle of the Scriptures meant to him to open up again the overwhelming source of living water, which was to be found in the oldest great theology, the theology of Paul, and in the tradition of Christ himself. The issue, as so often in a new paradigm, was a single thought that reorganized everything: justification by faith.

This thought is already so far removed from our age that one can hardly explain it to people today. (I know what difficulty I had as a young person in trying to understand what Luther was talking about; only after I had analyzed it, in the light of intellectual history, did I begin to have more than a sporadic empathy for him.) From the perspective of the Enlightenment paradigm of reason and of good will, the idea of justification by faith contains in fact two opposing components. On a superficial layer it is emancipating, and without this layer it would never have had such historical force. On the deeper layer, though, it denies precisely what the Enlightenment had believed in: the gift for moral and rational progress that is inherent in human nature. With the deeper layer I mean the deeper insight, even though the form of this knowledge, which was indispensable for Luther, does no longer correspond to our rationality. Although we know that it remains inadequate, let us try to find a modern language for this knowledge.

It is emancipating to realize that we do cannot derive our identity from the fulfillment of an externally given commandment. In this statement the concept of identity which itself is a result of emancipation stands for the pre-emancipatory concept of justification. The externally given commandment stands for what Luther, following St. Paul, calls the works of the law. But St. Paul was philosophically educated. He could, in Hellenistic version, think the thought that Kant has most clearly formulated in modern times: The law in its core is a law of reason that the mature human person, because of reasoned knowledge, is compelled to apply to himself. He who thinks in this manner, may say 'identity' instead of 'justification.' But St. Paul, and with him Luther, criticized exactly this central thought. Experience teaches me that I cannot do the good that I want, and even, if we look deeper, that I am unable to want the good that I affirm. This is not the place to consider how Kant, who saw this problem, tried to solve it; the Enlightenment, by and large, was not able to see the problem and has displaced it with the psychic instrument of optimism. The punishment for this displacement is the "dialectic of the Enlightenment", the always threatening collapse of the entire modern civilization.

Where the rationalism of the Enlightenment becomes aware of the problem, it usually blames the irrationality of the emotions. This conventional view of the human person, the analysis of the soul dividing it into a rational and an irrational part, is itself a 'level' of anthropological theory, a paradigm, that in our Western tradition stems from the Greek philosophy, from an interpretation (an imprecise one) of the parable of Plato. Cultural critique as well as the science of our time destroy this smooth solution and reveal its false elegance. On the one hand, cultural critique urges us to speak about the irrationality of the rational, of the destruction of the life-world through technical and political actions that are indeed rational if viewed from the perspective of particular interests and which the optimism of the Enlightenment considered reasonable. The minimum that must be accomplished here philosophically is the distinction between intelligence as the capacity to think conceptually and reason as the capacity to perceive the whole. But how can we gain access to reason? Science, on the other hand, specifically comparative behavioral science and the psychology instructed by it, teaches us the rationality of the irrational, the reason of the emotions. Emotions are perceptions without concepts; their disorder indicates only the disorder of the controlling self.

The experience that drove Luther to a monastery and then out of it, and which he found expressed in St. Paul, was, expressed in modern terms, that the ego is itself the origin of evil, not the emotions that are attached to it. Freud penetrated into this realm of experience, though with completely different preconceptions. Freud's relationship to Luther's problem can be indicated in three ways. First, no doubt, sexual problems had a key role for the young Luther, and thus for his entire life. One might see this in the face of the young ascetic as well as in that of the corpulent pater familias. It does not diminish Luther's greatness if one has realized, with Freud, the universality of the sexual problem. The frightening feelings of anxiety and the smoldering experience of sinfulness without apparent cause, Freud has taught us to understand as phenomena of sexual repression. Luther could never do justice to a monastic form of life because it symbolized to him a false solution of this problem. One may wonder if his later depressions were not connected to the realization that marriage, too, was not the full solution. Secondly, Freud taught that the ego should be seen not as master but as a relatively small organ of the psyche. However, Freud thereby descended into a labyrinth in which he too lost his way, especially because his conceptual tools, coming from the contemporary world view shaped by physics, were not suitable for his purposes. Thirdly, most importantly: As Freud recognized, the cure of a neurosis depends upon my acceptance of the acts of the unconscious as acts of my own will and on my accepting responsibility for them, even if I did not previously know that I had wanted them. This is the insight of Luther and of St. Paul: I am the one—not my emotions (how could I gain control of them?)—I myself am the one, who wants the evil that I must condemn. How that happens may be puzzling, but this is the experience. In the symbolic language of religion: There is forgiveness only for the repentant sinner, not for the one who does not admit to his own sins; and it is not even within his power to *want* repentance.

This experience of *servum arbitrium is,* in various cultural manifestations, the departure point of all religions that speak of salvation. In Buddhism and in Hinduism, despite all their differences, salvation is being experienced like an awakening. In important traditions, one speaks of an awakening to the true, divine self. In Christianity salvation is divine grace. Luther, who in accordance with his own personal experiences was especially mistrustful of every possessive claim of the self, also of the entitlement of grace or of divine participation, found redemption, i.e. the form of the relationship to God that he described as justification, in faith. The fact that his own experience could become exemplary and could have an impact on world history resonates with the inner core of the Christian civilization.

It is thus not the principle of scriptural authority that characterizes Luther's theological 'level', but what he discovered as the decisive content of the scriptures: faith. The opponent is not the pope but the principle of reason that prevailed from the scholastics to the Enlightenment. It might be Luther's heritage that Germany politically has never entirely consummated the Enlightenment, has never followed it uncritically. Despite modern efficiency, Germany remains a country of profound conservatives. Luther could not think of his experience in another rationality than that of the traditional interpretation of the Bible. No 'level' of the historical church or theology could be established on the basis of his experience as such. The Protestant church finally became the hostile brother of the Catholic church, or a promoter of the Enlightenment. Only in our own century is the Enlightenment affecting the entire earth, transforming all cultures. Its superiority to all traditions is obvious; its incapacity to solve its self-produced problems is becoming apparent and should not surprise us. Today, completely different spiritual 'levels' are necessary than all traditional 'levels' of Christian theology. But it will be a criterion for them whether they can teach us to comprehend what this traditional theology was all about.

Prof. Dr. Carl Friedrich von Weizsäcker (1975). *Source* The photo collection of the C.F. von Weizsäcker Society represented by Dr. Bruno Redeker who granted permission to use this photo. Original source unknown

Chapter 9
A Dialogue on Meditation (A Conversation with Udo Reiter)

Reiter[1]:

Professor von Weizsäcker, I would like to begin our conversation with a personal question: When you talk about meditation, you are not just dealing with theoretical reflections, but with your actual experience. May I ask what importance meditation, the practice of meditating, has had in your life?[2]

Weizsäcker:

Yes, you are asking an important question that I cannot really answer in a traditional manner. First, you are right: Meditation interests me mainly as something one does—not for any theoretical reasons. Personally, I have never had any formal training in meditation in the way one could get it and, if one were to be serious about it, should really have it. What happened was that as a youngster I simply began something that people, who are versed in this, told me later was meditation. So this was quite spontaneous in the beginning and really began in my childhood. Later I experienced in a Christian context how a liturgically shaped order of the day would look like. True, I never participated in this on a permanent basis, as member of a group, but I have occasionally been a guest of people who live in this manner. They were the Protestant St. Michael's Brotherhood, and from them I received a great present, namely, the knowledge that a structured way of life can make deeper inner layers accessible to us and can let these layers speak to us. I acquired the habit to keep silent for a while each morning, to keep still, and through this a lot made itself heard. Later, I also encountered Eastern kinds of

[1] The dialogue was taped, without preparation, for a radio broadcast, and was first published in: Udo Reiter, *Meditation—Wege zum Selbst* (Munich: Mosaik Verlag, 1976).

[2] This text was first published in English in: C.F. von Weizsäcker. *The Ambivalence of Progress*. Paragon House, New York 1988, 279–290; the German original is part of C.F. von Weizsäcker, *Der Garten des Menschlichen*. Hanser Verlag, München 1977, 533–550. All rights of the Weizsäcker Bequest were transferred to the Carl Friedrich von Weizsäcker Society, a nonprofit society whose executive chairman, Dr. Bruno Redeker, granted permission in March 2014 to use selected texts of Carl Friedrich von Weizsäcker.

K. Raiser (ed.), *Carl Friedrich von Weizsäcker: Major Texts on Religion*, SpringerBriefs on Pioneers in Science and Practice 24,
DOI: 10.1007/978-3-319-03704-2_9, © The Author(s) 2014

meditation, and it was an Indian meditation teacher who told me that my childhood habit was already a form of meditation.

Reiter:

To give our readers a proper understanding of this subject, let me ask right away the central question, the question about the nature of meditative experience. What really is it that is experienced in meditation? Is it possible to define it in terms of content? Can it be circumscribed through metaphors?

Weizsäcker:

What I should say immediately is that anything one could say about this is wrong, because this experience transcends the realm of concepts, the realm of what one can ordinarily convey in language. If we nevertheless deal with it through language, it depends on whom we talk to; either we talk to someone who has had these experiences, and then we understand each other almost without the use of words, or we talk to someone who has not had these experiences, and then the other one will find everything we say bizarre, or we find that our partner misinterprets what we say. This is one of the problems, but it is a problem that we find with any kind of experience, not only with meditation, but perhaps it is especially strong with meditation.

If you insist nevertheless on an answer—the kind of answer a psychologist might give—then I would say this: It is a quieting of the hustle and bustle of consciousness and then something, which has always existed, now appears. Generally, in meditation one does not become someone else, but one becomes the one that one has always been. But this becomes manifest in that, what we ordinarily call consciousness begins to become aware of it, and then the consciousness itself becomes changed through it.

Meditation, throughout history, has existed mainly in religion. We can, therefore, try to approach the matter with concepts of religious metaphysics. And what we experience here is, indeed, closely related to what we call 'prayer' in the religious language of the West; but if we say, it is an experience of God, then this is just as misleading as anything else we could say—except when we experience it. When we experience it, it is no longer misleading, but is so obvious that there is no need to communicate it.

Reiter:

Despite this difficulty, let me try again to talk about the substance of this experience. After all, meditative experience reaches a point that in our language is usually called an experience of enlightenment. You once described such an experience, and you wrote this sentence: "The knowledge was there and within half an hour everything had happened."[3] For clarification, let me ask bluntly: What

[3] Cf. In this volume Chap. 1 'Selfportrait', p. 23.

knowledge was there? This being an experience of wholeness, isn't it that, in Christian language, the individual feels suddenly safe and secure "in God"? Does this thought go in the right direction?

Weizsäcker:

You see I hesitate to answer immediately, and this hesitation is almost the most important part of the answer. I could say: When Christians report having experienced that they were safe and secure in God or have had a glimpse of God, then I am ready to believe they had such an experience. But that sentence of mine that you just quoted: I *wrote down*, that means, I considered well what I wanted to say, and I knew exactly why I wrote "the knowledge' but not *which* knowledge. The passage you referred to concerns something I experienced in India, at the grave of a great Indian sage, the Maharishi, in Tiruvanamallai. And I believe I will speak clearer when I remain apparently unclear and say: At that moment, standing in that place, in half a second, I knew: "I see!"—and this "I see!" is something that must be experienced. If one talks about it—and I have, after all, begun to talk about it, and I am ready to use all the words at my command to explain how this relates to our philosophy and our physics and our psychology—in short, if one talks about it, it will turn out that all these words are misleading, and with them we would probably mislead more than if we simply had remained silent.

Reiter:

Can we conclude from all this that meditation is not primarily concerned with some kind of content or even some new insight into the nature of the world or of existence, but that any content has only a subordinate role; that what counts in meditation is not the cultural background or the kind of knowledge one brings into meditation; that the Indian guru, the Japanese Zen master, the Christian monk, and perhaps even a German physicist meet somewhere at an ultimate point where content ceases to play a role?

Weizsäcker:

If I were compelled to answer the question of whether they do or do not meet with a mere 'yes' or 'no,' without being allowed to add another word, then I would say: Yes, they do meet. But if I am permitted to add something, then I say: This 'yes,' too, can only be understood if the meeting has really taken place, and this 'yes' is not at all meant to justify a relativistic or syncretistic denial of differences, where differences still exist.

For example, in the history of culture and religion we see that visions do exist but that, ordinarily, the member of a particular culture or religion encounters in these visions precisely those forms that are customary in the respective culture. And a skeptical psychologist would then say: Of course, it is obvious that what the individual sees are the subconscious cultural contents. And it is precisely from this that the skeptical psychologist would conclude that what the individual encounters is not what he thinks it is. What he sees is not the Holy Virgin or what he sees is not some corresponding Indian figure—what he sees is his own culture. But when I say—which is what I believe—that he nevertheless encounters reality or is "in the reality," then the way I must explain this to myself as a scientist is, that this being-in-the-reality happens to his consciousness with the means that are accessible to his consciousness because of its particular formation. Seen in this light, the differences should be taken seriously.

Reiter:

When we talk, here in the West, about 'reality' or about 'truth,' then we usually mean truth that can be proven empirically in scientific terms. Is it possible to explain how this Western scientific truth differs from the truth of meditation?

Weizsäcker:

I am constantly tempted to answer to your questions that one must not ask questions in this manner. But this is just a temptation, and I will not really yield to it.

First I would say: A fundamental experience of mysticism, a fundamental experience which is the aim of meditation and which is already approached in the most basic and the simplest steps of meditation, is the experience of unity, of the One. What 'one' is can ultimately not be asked, because a second thing would then be added, namely, the explanation of what it is. Also, the experience of unity makes it ultimately impossible to say how one can distinguish what is being studied in science from the experience of meditation; because then we would not have an experience of unity, we would have multiplicity. But if I enter the level of multiplicity, i.e. the level which science studies, then I can say in scientific language, in what sense science cannot express this unity—cannot express it as science—although it is the same reality. My personal opinion is that the reality that the physicist studies, the reality that the historian studies, the reality that the psychologist studies, perhaps even the reality that the mathematician studies is precisely the reality that can be experienced—perhaps—ultimately in meditation, because otherwise it would not be reality.

But, to get to the core of your question: we are able to say what our science can and what it cannot do. Science operates with concepts, and the concept is based on differentiation and subsequent comprehensive integration. When I say: "This animal is a cat," then I have distinguished it from dogs, birds, and everything else; but I have merged it with all cats. This method of using the concept to dissect the reality and then reintegrate it seems to me to be the basis of any scientific process, whereas the kind of exercise that we have come to call meditation is basically a training in a different mode of conduct, a conduct that does not proceed from dissection in order to reintegrate but that, I would like to say, proceeds from an acceptance of the non-dissected, and thus forgoes the achievement of integration; because achievement is precisely what is not the issue here.

Reiter:

You may have been critical of my way of asking questions, but I have the impression that the result shows, that I was on the right track, and with your permission I would like to continue in this vein. You said: In meditation one becomes *the one that one has always been.* If we view this statement in a positive light, then it is based, in my opinion, on the idea that in the innermost core of the human psyche there is something like a pure, genuine self that exists in itself in eternal identity, and which has merely to be cleansed of everything that has come about, of everything accidental, non-essential, to emerge as the core of the personality.

However, since Sigmund Freud, we have come to see things differently. It is difficult still to believe in this absolute core of the personality; rather, we tend to see in the innermost core of the human psyche a hard-to-define chaos of impulses and motivations, which overlap and reinforce or cancel each other—and the result of these motivations is then the very fragile and changeable product that we call the personality or the self of a human person.

Against this background, let me ask two questions: Are you nevertheless willing to maintain the fiction of an original self, "the one that one has always been"? And the second question (if you somehow share my misgivings): Is it really good and thus desirable to bring out into open—and I phrase it deliberately this way—*that which one has always been*?

Weizsäcker:

Well, I don't want to say at all that your question is not asked the right way. Earlier, too, I did not really say that your questioning was not right; what I said was that one must not ask in this manner. But one does ask questions, after all, and we have to live with this paradox, and we somehow manage to get by.

As for your question, it is very complex. First: concerning Freud. As you brought him into the picture, let me tell you what I think about Freud. I have not met him personally, but I have read him attentively and I have had, and still have, many close contacts with people who think in his tradition. It is my impression that Freud had a genius for seeing phenomena; that he saw phenomena that many others did not see. At the same time, he often was a somewhat stubborn adherent of certain scientific views of the nineteenth century. What is generally accepted of Freud today is, unfortunately, primarily that which is not good about him. What is good about him would be, learning to see in his manner. But it is so difficult to learn this that the people, who use Freudian concepts, notice very rarely that they are unable to do so. Therefore, I have the greatest respect for Freud and yet great reservations about citing him in support of an argument.

Let's apply this to your question: As for all these theories of ego and id and superego, I am willing to go along provided we say that, what Freud called the ego is a product and not the whole. I am inclined to say it is an organ. And if I may use the very common word 'psyche' to denote the entity whose organ the ego is—a usage that psychologists sometimes permit—then I would consequently say: The ego is a psychic organ, and this organ, like many other organs, is either good or not so good, but, at any rate, it is not the ultimate, it can be derived from something else. If I have said that one becomes the one that one has always been, then this means, in the language of a psychology that employs Freud's ego concept, that the total relativization of the ego becomes manifest in the experience that lies behind these words. The ego suddenly realizes that it is not absolute, and thus it becomes aware—vividly, not conceptually aware—that it is an organ of something much greater; and it knows that it has always only been this organ and nothing else, and that it does not at all have to defend its identity, which it used to defend with all force against the onslaught of chaos, against that which Freud calls the id, if it is content to be an organ and not the thing itself. And then it suddenly realizes that it has become that which it has always been, but now *knowingly*; and then the ego, in a way, is suddenly no longer merely the organ, because it is the organ that knows it is an organ, and it is this quite complicated state of affairs that I express by saying that it becomes that which it has always been. And this is a deliberately paradoxical statement, which, by the way, corresponds to a thought that we find somewhere in Western tradition. I believe we find it in Nietzsche, but then, Nietzsche is much more traditional than he pretends to be—at any rate, this thought is: Become the one that you are!

Reiter:

To this statement I would like to add a question on a somewhat different level. What you have just explained also means, seen from a somewhat psychological-mechanical perspective, that in meditation many things may emerge that one would not necessarily like to see coming out?

Weizsäcker:

Yes, this is something I failed to deal with, but as I said, your question is complex. First I would say: The well-known principle that one should meditate under the guidance of a guru, to which we have a Western analogy in Freud's strict requirement that only a person who has been analyzed is allowed to analyze, this principle, this practical rule exists because in meditation, if we take it seriously, things can happen that destroy the personality, things that can take us along a path of which we later would say: I wish I had not taken this path!

One can convey the same by saying that out of the subconscious or, whatever we prefer to call it, out of the realm of the id or of the demons, there emerge contents and behavior patterns that are ordinarily, and justifiably, suppressed, and which a truly wise person may want to face, so that such a truly wise person would let all this emerge; but anyone, who does not have this true wisdom and does not have a teacher who could save him, is in danger of not being able to exorcise the spirits he has conjured; that what has emerged does not let him see the ego as the safe organ of the greater self, but only as a tiny boat on high sea, no longer capable of resistance. I have learned this, for example, from Indians like Gopi Krishna, whom I came to know quite well personally, and who described his experiences in his first book, *Kundalini*, and if one would show this to a Western psychiatrist, he would say: This, of course, is mental illness. At the same time, however, this is the Kundalini experience, because, as Gopi Krishna says himself in his present state of mind, a good portion of what we call "mentally ill" is of course such an experience that has turned out badly.

Reiter:

You describe dangers that are possibly connected with meditation; thus the simplistic equation "meditation = the good" must be seen with caution. In this context I would like to bring up a phenomenon that I have encountered in many devotees of meditation, namely, an almost limitless optimism; it ranges from the possibility of one's inner development to the lowering of the crime rate of a society, to evolutionary progress that is supposed to manifest itself even biologically (if we think, for example, of Aurobindo). Is it that such expectations are merely the customary sectarian hopes of salvation as we have seen them throughout the centuries, or do you think they have a solid core?

Weizsäcker:

I am ready to see a solid core in them. But if I am to talk about this further, I must first admit that much of what presents itself in the form of this optimism seems to me to be more or less naive, especially if we disregard here the further development of the human species and look simply at the individual. I can understand very well that people who have meditated, who have experienced what is beneficial in meditation, are so to speak inundated by it and suddenly see it as the sole and supreme fulfillment. From the perspective of the individual's life this is understandable and to a certain extent perhaps justified. Also, it is possible that an individual, who has had such experiences, progresses constantly and truly becomes a sage, a knower, along the path that you have called 'optimistic.' But I have the feeling that this could happen only in a cultural setting, in which what he experiences is considered of highest value; and even then one must assume that the advanced experiences that are made along this path are immensely complex and difficult.

So this voyage is not easy, and therefore I think that what I have just called 'inundating,' the inundation with this experience, will usually turn out to be a naive preliminary stage. This applies certainly to our Western culture, and I assume it also applies to Indian or Japanese culture if understood in an appropriately broad sense. In the real world, one must remain open to science and to plain reason and to common sense if one wants to demonstrate that what one has experienced is not merely regression—psychologically speaking. Maturity, after all, shows itself not only in having certain experiences more and more often, but also in being able to cope with the problems of our fellow humans and the world.

But after this preliminary remark that we are certainly dealing here with many dreams of the newly converted, I would say: I do, indeed, believe that this road makes experiences accessible that generally have been accessible only to few people in the past. A large part of religion seems to be merely the acceptance of these experiences by those who have not had them personally, but who were sufficiently impressed by them to understand that they represent something real.

I could imagine that these experiences are now becoming accessible to a larger number of people than in past ages, and that this may also contribute to a change of society and perhaps of human character in general. But I would not want to see this tied in with proclamations of grandiose visions for the future. I always feel very keenly: we live, think and philosophize in the here and now and the future is not within our reach, even though it might occasionally seem to reveal itself in a parable that we understand only after it has come into effect.

I would rather reverse the argument: As you know, I have been very actively involved with the problems of the prevention of war and with economic development, including the perspective of overcoming inequalities and poverty or preventing environmental destruction, or whatever the problems are that are in the background of short-term decisions, although the problems themselves are entirely concrete. And this regard I can only share the view of those people who today are called critics of our system, namely the opinion that the current system of the modern world is not solving these problems. This I say not specifically with regard to the capitalist system; the systems that call themselves socialist are, I believe, merely a new edition of the same mistakes.

And if I now ask: Why don't they solve these problems?—then I would answer: certainly not, because the problems are unsolvable. One could say with plain reason, even with common sense, what would have to be done to solve them. So what is needed is not a particularly intelligent person, who can solve complex matters of management and control; this is not the issue; in the last analysis, we realize that our inner constitution is such that each one of us, at one place or another, and some of us at many places, reject the only thing that is salutary, because each one is afraid of what might happen to him, if he did make this concession, which, however, is the only one that would save him.

And this structure of anxious self-protection of the ego can be overcome, if the ego does not see itself as the final reality that must be protected at any price, but—as I have said before—as an organ. And therefore I believe that, wherever the experience that we talked about becomes real, there is a chance that human beings cease to act in the fearful and self-destructive way that we see almost everywhere today. In this sense I could agree with the people who have those great hopes; but first we must get that far.

Reiter:

This is really the old question about human changeability. After 2000 years of Christianity, one has to look at this prospect with a certain skepticism. But this is something that you yourself have done indirectly?

Weizsäcker:

Yes, but here, too, I would prefer to reverse the argument. What I mean is this: Suppose, someone happens to read the Sermon on the Mount. I did this when I was a child; I read the Sermon on the Mount when I was eleven, and I felt immediately: If it is true what it says here, then all of our life is wrong, even the life of those who were close to me and whom I adored. I soon learned that it is almost impossible for humans to change this life. And yet, I never thought for a moment that the Sermon on the Mount could be wrong. This means: When Jesus spoke those words, he must have known that people were not going to do this. If you read the New Testament with this in mind, or also the prophets in the Old Testament, then you find many passages from which this becomes entirely clear.

But one does not refute Christianity by showing that people have not acted like this in reality. The text of the Sermon on the Mount could not have been written, if there had not been someone who has really lived this, who wrote out of his real experience, not out of some abstract demand.

This means: It is not true that the human being cannot change at all; one can. But this is possible—to use Christian language—not through one's own good deeds, but through the encounter with grace. But this grace, to speak again in Christian terms, is bestowed only to one who has shown through his own efforts that he is sincerely searching. To be sure, there is also grace beyond all reason; but in general grace is experienced as a response.

This is the second aspect. The third aspect is that I would simply dispute those who say that Christianity has not changed the world. On the contrary, I think Christianity has changed the world very rapidly. Animal species change over the course of hundreds of thousands of years; 6000 years of history are just a fraction of a second in comparison with the preceding history of the world. If humanity can change in such a way that something new happens approximately every thousand years, then this is tremendously fast, compared with animal life. The reason for this is that humanity can accumulate experiences, that it has language, that it can create tradition, that it can grapple with the past and the present, that it has reflective thought. And considering the pace that can possibly be expected from historical changes, I maintain that Christianity has changed the world greatly.

It has been my personal opinion, for a long time, that modern culture is largely a secularization of Christian contents, and I am using the word 'secularization' the way it was used by the Protestant theologian Gogarten, namely, as realization of that which was merely demanded before. But secularization always contains the danger that one forgets something else that was related to the demand. For example, social justice: This was one of the demands of Christianity, a demand that was not fulfilled in the ancient world, and which was not fulfilled in the early Christian world, but to which—as we can say optimistically—we have come closer and closer. But the demand for social justice does not release us from the other demand, the demand of charity, of love for our neighbor. This is something quite different…

Reiter:

… to which we have not gotten any closer?

Weizsäcker:

Oh yes, we are coming closer to it, too. But we would move away from it, if we thought that we could eliminate the people, whom we see as the obstacle for the demands of social justice. Isn't it that the essence of charity is also expressed in the sentence "We are all sinners"? But the people who work on social progress, while doing something laudable, are often of the opinion that only the others, the political opponents, are sinners. And whoever does not know that he himself is a sinner, will commit crimes. He cannot be held back. Self-righteousness is the root of all really serious crimes, and in the secularization of Christian contents I see the danger that one may forget this, and whoever forgets it will destroy everything. "We are all sinners, and we lack the righteousness that we should have"—to put it again in Christian terms. You could also express the same thought in another language, even in atheistic language, as Buddhism seems to do. I am not insisting here on referring to the Christian tradition. But I express myself in this tradition, because it is the one, in which I grew up and in which I express myself with ease.

Reiter:

I would like to ask another question about a concrete aspect of meditation. For a long time, there had been little talk about meditative experiences in the Western world. But this has changed radically a few years ago. We have witnessed a sudden interest in the teaching of meditation, above all among young people; this has looked for a while, and still looks, like a boom. What do you think are the reasons for this sudden yearning?

Weizsäcker:

It seems to me that once in a while a new reality is being discovered, probably at those times, when it is urgently needed. Some aspects of this boom may have taken on absurd dimensions and may have given opportunities to people who do not deserve them, but basically I consider this boom as an awakening, an urgent need.

I welcome it, because I think that the correct insight that we must make progress with regard to the social structure and the order of society becomes self-destructive, if it is not combined with those experiences which can be furthered by successful meditation. And I have the feeling that the disappointed social critics are the ones, who would benefit most from having the meditative experience.

Reiter:

This may be related to the observation that our scientific way of approaching solutions has led us into a crisis that almost threatens the existence of humanity today. Here, meditation seems to be seen as some kind of fundamental alternative. Do you share this impression?

Weizsäcker:

This, again, is a very interesting question. I tend to believe that the kind of science, that destroys the world, is bad science. This means: The dissecting and reintegrating of concepts, which I mentioned before, is of course correlated to a dissecting of the real world, to a physical dissection, a destruction of something that perhaps can never be reconstructed, where unity is lost forever. In this respect, science is certainly full of dangers, as we can see in many examples.

There are many people today, who for this very reason are strictly anti-scientific. But I think, in any such instance one could demonstrate scientifically that the 'dissecting' has taken place in a way, which has overlooked a part of reality that is still accessible scientifically. So I really believe that the issue is not primarily to replace science with something else, but to bring science to its proper level.

But I think, from a historical point of view, that the scientific schooling in conceptual thinking, which has been developed mainly in the West, but also in the East, and the schooling in meditation, which has been developed mainly in the East, but also in the West, are alternative paths. What they have in common is that they represent forms of schooling; what is alternative is their aim. And I believe: In an ultimate kind of reality, in the unity, of which we get a glimpse in meditation, both lead exactly to the same goal.

Reiter:

To look at the actual situation, here and now: A young person, who feels the need for meditation, finds the scene that you have described, but also sees some odd, exotic Asian practices, and also commercial interests. What advice do you have for this person?

Weizsäcker:

Although I am afraid I will get a lot of mail on this, which I will not have the time to answer, my answer would be that I would look at each person individually and perhaps give each one a different answer. I cannot give a general advice to meditate. Nor can I simply say not to do it. No guru can do this either; he can only look at the individual, and only after understanding the particular need can he say to follow this or that path.

The Western belief that one can give generally valid advice, a belief that is related to bad science, is wrong—like the advice that everyone should do this or that, for example, that everyone should use a particular toothpaste.

But there is one thing that I would say in a general way: One should not indiscriminately or in a surge of enthusiasm respond to a charismatic figure and become flooded by feelings that remain unexamined. One should never abandon one's alertness, one's rational consciousness, one's intellect in this situation. I have seen enough accidents happen when people have not headed this demand.

Reiter:

I would like to go back to one aspect that you briefly mentioned earlier. Opponents of meditation often maintain that meditation is a flight from reality, a withdrawal from the demands of this world, from human requirements, a retreat to one's own inner world. How would you counter this kind of criticism?

Weizsäcker:

I would like to give the same answer that I once heard from Gopi Krishna, whom I mentioned before. This was at my institute, when a young man asked practically the same question. Gopi Krishna looked at him in a very friendly way and said: "The more I hear Europeans talk about meditation, the more I feel like advising them against it. They just don't understand what it is all about. Read your own Holy Scriptures, and you find the same as in ours: Love your neighbor; love God; love your neighbor in God. And the rest does not count. Nowhere does it say that you should meditate. But if you want to love God and your neighbor and discover as a great truth that meditation can help you in this and may be decisive in achieving it, then you should meditate, and if you do not make the discovery of this need, then you should not meditate."

So this is not flight into one's own inner world, but it means facing those innermost inhibitions that prevent us from turning to our neighbor and to reality. And there is something else: A large part of what is thought to be an active engagement with reality is actually nothing but a flight from the truly looking at oneself.

Reiter:

I began our conversation with a personal question, and I would like to end it with one. Professor von Weizsäcker, it is not at all common today, in this kind of society, that a physicist, a man of the exact sciences, treasures meditative truth side by side with scientific truth. And, as far as I know, you have been attacked for this, or dismissed condescendingly as a guru or mystic. What has been your reaction? And what was the experience that made of the physicist Weizsäcker an adherent of meditation?

Weizsäcker:

Well, let me say this: There will always be criticism; unfortunately I do not often get to hear it; but this doesn't really matter.

Even as a child I was in love with science, and at the same time I lived in a way that I later learned had a meditative basis. When I was twelve years old, I was already very much occupied with the problem that these two realities had to be the same reality, but that I just did not have access to the intellectual means to comprehend this. With this extended background, I would say today that the unity of nature, which we have learned to see through the development of science, is basically a mirror image of the unity that we seek in meditation. So my answer has to be that I don't see a difference. I see a difference in methods, in cultural traditions, in the concepts we employ, but I do not believe there is a difference in substance.

Prof. Dr. Carl Friedrich von Weizsäcker (1981). © The Weizsäcker Family represented by Dr. Elisabeth Raiser who granted permission to use this photo

Chapter 10
What Is Meditation?

The three levels of the theoretically true, the aesthetically beautiful and the morally just are three forms of isolating one of the aspects of perception. Each of them signifies a powerful expansion of consciousness.[1] Each one of them has an immanent tendency toward autonomy, in which it is bound to fail, because it isolates only one segment of the true. I have elsewhere tried to indicate this for the core process of the theoretically true, for reflection. The same could be shown for the other two levels; I do not intend to do this here. There must be a third, an integrating level of truth above that of the factually true and the theoretically true. Such a level does not exist today among the products of modern occidental culture. In the history of humanity this place was so far occupied by religion. Among the above-listed levels, it was called by the name of the sacred. However, traditional religion is no longer in conformity with modern consciousness. It is just as futile to wish that this fact could be denied, as it would be to conclude from the acknowledgment of this fact a rejection either of religion or of modern consciousness. The real challenge imposing itself with intrinsic necessity is to achieve a transformation of the consciousness which would facilitate the rethinking of religion and integrating the disintegrated consciousness. Philosophy alone cannot generate this real process, but it can work along with it.

The modern consciousness must ask: What is religion? Again, and without any prior systematic claim, I have tried to distinguish four aspects from among the abundance of phenomena that can be subsumed under the label religion: Religion as the bearer of a culture; religion as foundation of a radical ethics; religion as inner experience; and religion as a theology.[2] In retrospect these distinctions seems to correspond to the aspects of perception.

[1] This text has been translated from Chap. I, 10.5 "Meditation und Wahrnehmung", in: C.F. von Weizsäcker *Zeit und Wissen*, Hanser Verlag, München 1992, 499 502. All rights of the Weizsäcker Bequest were transferred to the Carl Friedrich von Weizsäcker Society, a nonprofit society whose executive chairman, Dr. Bruno Redeker, granted permission in March 2014 to use selected texts of Carl Friedrich von Weizsäcker.

[2] Cf. in this volume Chap. 6 "Thoughts of a Non-Theologian on the Theological Development of Dietrich Bonhoeffer", p. 86 ff.

K. Raiser (ed.), *Carl Friedrich von Weizsäcker: Major Texts on Religion*, SpringerBriefs on Pioneers in Science and Practice 24,
DOI: 10.1007/978-3-319-03704-2_10, © The Author(s) 2014

Religion as the bearer of a culture shapes social life, structures the times, determines or justifies morality, interprets anxieties, shapes joys, consoles the helpless, gives meaning to the world.[3] It is the way in which a whole—that is, a culture, the life of a human society in the world—understands itself. However, this form of wholeness of human life becomes something identifiable for us only, once it has slipped away from us; once it has become historical to the observer, even if it still were continuing. The other three aspects of religion are attempts to preserve the substance of religion by separating it from the nonsubstantial.

Theology is judgmental thinking. The designation 'theology' first denoted the core of Greek philosophy. In a language itself originating from philosophy one could say of Western theology: It addressed the whole of immanent life in the world by attempting to express transcendence in the form of judgements. Religion as the basis of a radical ethics has not only given rise historically to the level of the morally just. It certainly achieved this through the distinction between good and evil, and symbolically in the Ten Commandments. However, the prophets of the Old Testament and the Sermon on the Mount go beyond that level. They also speak of healing the immanent ambivalence of morality.[4] Morality divides, love binds. Here, too, we encounter the inherent relatedness between the aspects of perception: We count morality among rules of conduct, love among the emotions. Here too, transcendence comes in: expressed in Christian terms, we can only love our neighbour in God. One might, finally, be inclined to assign religion as inner experience to the emotions. Meditation is counted among these. But we see immediately that meditation is something entirely different, a unity of perception through denial of all its traditional aspects.

How does meditation enter into our experience? It begins with a spontaneous denial of that which constantly imposes itself upon us as being the real. When asked in conversation what the meditative experience is, one can initially only remain silent, if one really wants to indicate it.[5] Remaining silent *is* the beginning of the answer. Where meditation is properly practiced as an art, denial is a well-considered technical tool.

Action is denied. The meditating person performs no action, he does not speak; while he is meditating, he can be in communion only with those who meditate, a profound, silent community—from the Quaker meeting to the meditation hall of Hindus and Buddhists. The space of a Christian church, all liturgical rituals do in fact also exclude all other forms of action apart from the inner, religious act.

Sensory perception is denied. Often a closed space is sought. Bodhidharma "sat facing the wall for 9 years." But even someone capable of meditating in the

---

[3] *Loc.cit* p. 86.

[4] Cf. in this volume Chap. 10 "The Moral Problem.", p. 119 ff, as well as Chap. 6, *loc.cit* p. 87 f.

[5] Cf. in this volume Chap. 9: "A Dialogue on Meditation", p. 100 f.

open air, as the teachings of the Buddha describe: “he seeks a lonely spot, on the plain, in the depths of the woods”; he thus initially escapes from the sensory stimuli of the world of human beings, and soon the nature surrounding him is no longer a stimulus but a silent partner.

Judgmental thinking is denied. “The act of thinking is aimless” (M. Steinke). Thoughts are the monkies to be tied down. The paradox, the koan, breaks the familiar line of consequence of all thinking; it interrupts the stream of logic.

The emotions are calmed, acknowledged, put aside. The multifarious forms of happiness, the graspings for bliss prove to be just as many obstacles to bliss. “The sea’s inner calm” is to be attained.

Now, what is perceived when the separate aspects of perception remain silent? Let us begin with what a master can still describe in easily understandable words to a pupil; let us begin with a meditation on an object, for instance, meditation on a flower. You look at the flower and at nothing else; you close your eyes and you still see it. This simple meditation uses the separation of aspects in a somewhat different form than in the case of reflection. Action is cut off when you have sat down for viewing. Judgmental thinking may come to rest. But predicative perception remains there with full integrity. You do not see spots of color making you judge: ‘flower.’ You see the flower. As the emotions calm down, emotiveness attains its purity: a not-covetous, quiet delight in the flower. Predicative, emotive perception is reduced to its essence and experienced holistically. You have closed your eyes. The real process of sensory perception is now also interrupted: The essence of the perception remains unchanged to you. Verbal reflection is excluded. What do you actually see in the persisting image? Do you see this flower or the eidos? You see neither of these two products of reflection. You see this flower as a flower in its integrity. By cutting off what does not belong to it, you experience the perception in its unreflected wholeness. You experience the unity of the perceived with the co-perceived.

From here on, the description in judgmental speech necessarily gets more and more difficult. For judgmental speech is based on the sundering of what here remains unsundered and on the forgetting of what is being perceived here. A few suggestive sentences may be allowed. You learn to perceive not the flower but the perception; not the perception but the co-perceiving self. You no longer experience the flower, no longer a consciousness of the flower, no longer a consciousness of the self, of consciousness—you are consciousness.

Chapter 11
The Moral Problem of the Left and the Moral Problem of Morality

11.1 The Moralization of Politics

We are witnessing a politicization of humankind. Never before has such a large percentage of people been concerned with politics. This is an aspect of the democratization of politics.[1]

This process has also led directly to a moralization of politics. Politics is being perceived less as the game of the big shots, less as the business of professionals, and perceived less as fate. On the contrary, it is the topic of morally judged decisions in which everyone is called to participate. Whether in this way politics has become more moral is certainly doubtful. Certainly it is becoming more moralizing. The appeal to moral judgments has become more and more an indispensable means of politics. The very ancient tactic of lying and the inclination to affirm oneself through self-deceit, that were always an aspect of politics, today present themselves more and more as an ideology through reference to universally valid, moral principles. The moralization of politics is an example for the ambivalence of progress.

The following comment discusses this issue on the basis of a concrete example. A specific incident has evoked in me once again the emotions that I describe at the beginning of my comment. The writing resulted from the attempt to objectify these emotions by analyzing their origins, in order to master them. This comment, because of its emotionally motivated origins, mirrors in a much franker way my true feelings about the phenomenon of "the Left," much more than my other, more controlled writings. At the same time, however, it touches on problems that in my

[1] This text, which is a private note written in 1975, was first published in English in: C.F. von Weizsäcker: The Ambivalence of Progress. (New York: Paragon House, 1988). 84ff; the German original is part of C.F. von Weizsäcker Der Garten des Menschlichen. (München: Hanser Verlag, 1977): 116–121. All rights of the Weizsäcker Bequest were transferred to the Carl Friedrich von Weizsäcker Society, a nonprofit society whose executive chairman, Dr. Bruno Redeker, granted permission in March 2014 to use selected texts of Carl Friedrich von Weizsäcker.

K. Raiser (ed.), *Carl Friedrich von Weizsäcker: Major Texts on Religion*,
SpringerBriefs on Pioneers in Science and Practice 24,
DOI: 10.1007/978-3-319-03704-2_11, © The Author(s) 2014

opinion have central significance, far beyond the history of the leftist movement in recent years. Because of these problems, I wanted to put this comment into the hands of friends, particularly those who, in their basic attitude, are much closer to the Left than I am. But this comment is not so much concerned with those emotions, but rather with the question of how to overcome them by seeking to clarify the reasons for the ambivalence of behavior that wants to promote progress.

11.2 Comment

This reflection moves from the particular to questions of principle. It begins with an example taken from the field of experience which the writer shares with others in his profession or of his generation.

The words "the Left" or 'leftists' are first of all used to designate the political tendency that was dominant in the student movement in the 1960s and early 1970s in countries like our own. Liberal professors (but also liberal politicians, judges etc.) have often had the following experience with these leftists: The professor concerned was a critic of many of the structures of the society in which he lived. He had greeted the beginnings of the student movement with sympathy and with great willingness to learn. He saw the intellectual force of the mostly Marxist-inspired global interpretation of society and its importance as a political-economic perspective. He was fascinated by the determined commitment, and this not without self-criticism, with which he reflected on his own factual, bourgeois conformism. In the last analysis, his sympathy with the Left was not based on theoretical agreement—after all, he found the leftist theories too confused—but on the deep impression that the young people made upon him with their *moral* motivation. He easily overcame his first bourgeois shock caused by loose manners, sexual freedom, and similar breaches of traditional morality, because even in these breaches, in which he never participated personally, he sensed the moral nature of the need for truthfulness, he sensed the morally motivated protest against the moral hypocrisy of the prevalent external forms. He offered the leftists open conversation, free cooperation, and protection against repression by the dominant system. Not in every case, but in a significant number of cases, he experienced sooner or later that his trust was crudely and irreparably being misused and abused. He came to see himself confronted by a conscious plot to seize power and realized that he had only been of interest to the leftists, as long as they could use him.

He understood that he had been a "useful idiot". This explains the presently unhealable trauma from which many of those, who had been sincerely progressive liberals, suffer in relation to the Left. It often causes the previously progressive persons to assume a reactionary stance that they themselves would have criticized sharply a decade earlier. I assume, too, that this is the deepest reason why, at least for the time being, the leftist movement has radically and miserably failed in all highly industrialized countries with representative democracies. While the criticisms of the system of domination, which exists in this society behind the formal rules of the state of law, are certainly justified, the Left still could not win ground over against the simple, but not easily deceived *moral* judgment of the non-intellectuals, and particularly the workers, to whom the left frequently refers. The Left has lost out so far over against a system that is indeed morally questionable and has legitimately been challenged by it, because its own factual morality has induced a moral scare, which, even while clumsily expressed, was fully justified. The morally highly motivated Left thus fails precisely because of its systematic violation of morality.

It is obvious that my analysis is highly controversial in its interpretation of the facts. I would be prepared to enter into a discussion as to its correctness. In order to control my perceptions, I would further be prepared to present more details in support of my thesis regarding our own and other societies. In my present comment I start from the assumption that this analysis has captured correctly at least one dimension of the events. This gives rise to the question of how developments like these can be explained. The moral problem of the Left then leads on to the moral problem of morality.

We are concerned here first of all with the relationship between morality and society. When I speak here in a somewhat abbreviated way of morality, I am thinking of the highest form of moral principles so far developed, of the universalistic morality. It has found its time-honored formulation in the golden rule, "Do not impose on others what you do not wish for yourself"; its most thoroughly reflected philosophical expression is represented by Kant's categorical imperative: "Act only in accordance with that maxim through which you can at the same time will that it become a universal law." There has existed in human history, for thousands of years, the phenomenon of domination, that is to say: a manifest inequality of socially guaranteed rights. Traditional morality that in most cases was rooted in religion has found compromises between the recognition of this fact and the universalistic morality. This includes the ethos of higher obligations on the part of the rulers. Its great actual significance is being disregarded most of the time by the contemporary leftist critics to an extent that indicates a fading of reality. Another way out is the renunciation of participation in the system of dominance and material wealth by persons who live as hermits and mendicant friars or in sects. The ethos of nobility in its core, and obviously also the ethos of renunciation, are both supported by the conviction that the demand to treat other human persons equitably can only be fulfilled, if I demand more of myself than what I expect from others. These forms of ethics have been possible by virtue of their religious core: Not the self-accomplished individual righteousness—which is never attained—is the basis of moral behavior, but divine grace, which fills the gaps that are being left, even by the most well-intentioned ways of acting. This religious experience is just as real as the un-contradictable validity of the universalistic morality.

The tradition of the radical European Enlightenment, which the contemporary Left is following, attacks the very fact of domination. It tends towards the abolition of domination. I will not comment on whether this will someday be possible in a radical sense. On the basis of my assessment of anthropological conditions, I would have to consider it to be possible in principle, but only in a future that will remain unattainable for a long time. I address here the problems that arise if one hopes and tries to realize this directly, that is to say in one go. The leftists, who hope to accomplish it either through an act of revolution or in the still unbelievably short process of a single "long march through the institutions", criticize directly the institution of domination itself from the perspective of morality. They call this the demand for justice. They see through and criticize the conduct of the rulers who invoke the ethos of nobility largely where it does not hurt their own rule.

This leftist criticism now encounters the old moral problem of ends and means. It recognizes the social conditioning and the largely concealed ('ideological') purpose of moral judgments. It is convinced that only a change of society that would abolish domination as the actual precondition for the moral lie would make possible a genuine universalistic morality in society. Its adherents, therefore, feel legitimated to use a non-equitable morality against the representatives of the existing system, i.e. to treat them in a manner, in which they would not like to be treated themselves. They repress the truth that they betray the morality, which they want to establish, on the way to its establishment by their ways of acting, and that a somewhat sensitive person cannot fail to see the betrayal. So they create the situation where their own morality is being discredited. But once this insight takes root, they then find themselves in a desperate situation; because they known that the dominant system cannot be overthrown by other means than those which they try to employ. If these means fail, the system cannot be overthrown in the foreseeable future.

I am not talking about what might happen to the system in the long run, but rather about the moral problem of the Left. In the following sense it is the moral problem of morality itself. Morality in a reasonably radical sense is possible, if one renounces social security, as in the case of the above-mentioned religious groups. It should, however, be a genuine obligation for people who reflect responsibly about politics, to contribute to the creation of such social conditions that enable ordinary people, who are not radical nonconformists, to act as best as possible in accordance with the demands of morality. What do we say, when realizing this aim against the existing powers would be possible only by violating moral principles?

I maintain that this problem allows for many practical solutions when individual situations are considered, but that a fundamental solution only on the basis of *plain morality* cannot exist By the concept of plain morality I understand a morality that might be based on the golden rule or the categorical imperative, but which is not rooted more deeply in what I have called the religious experience. This is the experience of grace, of the redeeming power of loving one's neighbor, and, ultimately, in the love, honor, and fear of that deepest self that is called God in the religious tradition. Without this experience, there is no bridge between unrealizable rigor and foul compromises. Both of these positions lead a morally sensitive person to self-hate and, through the psychological mechanism of projection, to hatred of others. This hatred is the real basis for the moral failure of the Left. I believe that one can discern in this line of thought the 'dialectic' of leftist morality: Precisely *because* the Left is first of all morally motivated, it lapses into deeper moral mistakes, than its morally less active opponents. Therefore, I deliberately abstain from moral condemnation of these moral mistakes; they are basically a phenomenon of desperation. But they have those suicidal consequences, which inhere in all products of desperation. Those, who never have experienced these temptations, have no reason for moral self-satisfaction. The genuinely productive path does not end with this desperation, but begins when we gain courage to confront it directly. One can reduce the moral problem of morality to a simple formula of which intellectuals would normally be ashamed, because of its simplicity: The ultimate basis of human cohabitation is love and not morality. Morality is a penultimate basis.

Prof. Dr. Carl Friedrich von Weizsäcker meeting with the Pope John Paul II in Gandolfo (1985). *Source* Vatican/Osservatore Romano who granted permission to use this photo

Chapter 12
Notes on the Relationship of Physics and Religion

The general position of contemporary physicists concerning religion appears to be agnostic, but open. Conscious antireligious convictions by physicists occur less seldom than with biologists, and much less seldom than with sociologists. This might stem from the fact that the dispute between physics and the church lies farthest back in history. It remains to be seen what this historical succession might mean.[1]

Freud speaks (in *One Difficulty of Psychoanalysis*, 1917) of the three offenses against human self-consciousness committed by Copernicus, Darwin, and psychoanalysis. We had to learn that our planet is not the center of the universe, that the other creatures are our brothers, and that "the ego is not master of the house." The third offense is the deepest. If we separate this observation from its special connection with psychoanalysis, then it can be said that the sobering self-awareness through science continually advances closer to the human personality core.

But one must see that this continuing disillusionment means continuing conflict with religion only, if one thinks of religion, from the beginning, as representative of the glorification of the human ego. On the other hand, one could also see in the disillusionment about the ego an approximation to the core of religious experience. Should not religion make us humble? But the contents of this scientific disillusionment have not, so far, been religious, and the ego finds compensation for the offense in the fact that it itself has become the bearer of knowledge.

But physics has a doubly positive access, at least to the question of religion. The one access might be exemplified by people like Kepler. In the mathematical laws of nature an unforeseeable majesty is revealed. Kepler interpreted this experience,

---

[1] This text, which was written in 1976 as a contribution to an interdisciplinary seminar at the University of Munich, was first published in English in: C.F. von Weizsäcker: *The Ambivalence of Progress*. (New York: Paragon House, 1988) 255f; the German original is part of C.F. von Weizsäcker: *Der Garten des Menschlichen*. (München: Hanser Verlag, 1977): 441–443. All rights of the Weizsäcker Bequest were transferred to the Carl Friedrich von Weizsäcker Society, a nonprofit society whose executive chairman, Dr. Bruno Redeker, granted permission in March 2014 to use selected texts of Carl Friedrich von Weizsäcker.

K. Raiser (ed.), *Carl Friedrich von Weizsäcker: Major Texts on Religion*,
SpringerBriefs on Pioneers in Science and Practice 24,
DOI: 10.1007/978-3-319-03704-2_12, © The Author(s) 2014

one shared by many physicists, through Christian Platonism. The world is created in accordance with the creative thoughts of God, which means that it has been constructed in mathematical harmony. The human being, created in the image of God, can imitate these thoughts. Science is worship of God.

The other access was opened by the physics of the first third of our century. The image of the world of classical physics was objectivistic, deterministic; consciousness was outside of nature thus described. The quantum theory can only be interpreted, if one takes into explicit consideration the relation between the objectivizing description of nature and the observer. Thus the primitive ontology of classical science was shaken. Philosophical questions have been asked that still have no answer.

The conclusion of the dispute between physics and the church (the "revision of the Galileo verdict") does not appear to be the most important aspect of today's situation. Scientists and Christians can do each other a service if they continue to ask each other critical questions.

Christians must ask the scientists whether what they do to the world is not, perhaps, criminal (destruction of the world through the consequences of objective knowledge). Questions, not accusations. Only self-accusation opens the sources of grace; one can and may defend oneself against the accusation of others. Objectively, not subjectively. The redeeming experience of sin begins where we learn to identify with our actions, although the subjective intention was perhaps not evil. A science that went through this experience might find itself. A science shielded from this experience becomes objectively evil.

Scientists must ask Christians whether they have consummated the modern consciousness. Much of this consummation has been caught up within Protestant theology in the last decades. Here modern consciousness had often been too simply accepted as truth. But as deconstruction of an untenable (and fear-conditioned) self-assertion of the church, this process was necessary. Perhaps, what has been said of the necessary experience of science, applies, *mutatis mutandis*, also to the necessary experience of the church. It is neither a disgrace nor dangerous to admit that the intellectual problems between religious truth and modern consciousness remain unresolved.

Prof. Dr. Carl Friedrich von Weizsäcker meeting with the Pope John Paul II in Gandolfo (1985). *Source* Vatican/Osservatore Romano who granted permission to use this photo. Original source unknown

Chapter 13
Unfinished Religion

We are concerned about a change of consciousness. Political consciousness, i.e. commonly applied political reason is pre-conditioned by individual consciousness; hence a transformation of public political consciousness presupposes a transformation of consciousness on the individual level. We encounter this change of consciousness in four forms: education, maturation, healing, and conversion. Education and maturation can each be called an *evolutive* change of consciousness; healing and conversion, a *radical* change of consciousness. The personal consciousness has a common, and in this sense public foundation in culture. We are therefore concerned about a cultural change of consciousness, a transformation of culture.[1]

Our line of reasoning initially has a pragmatic aspect. Education can be planned. If it is planned reasonably, it will conform to the processes of maturation one may expect in any healthy development. In this evolutive context one can talk of a political consciousness *growing up*. Such education must be presented with an image of *reasonable hope*. It is the hope that humankind will become viable and sustainable. That involves a step in political enlightenment.

Our line of reasoning, however, cannot stay with this pragmatism. The preconditions of our pragmatic actions are not under our control: "Consciousness is an unconscious act." The historical role of religion has been to take seriously the unconscious preconditions of our consciousness. In this regard it repeatedly saw itself forced to demand some form of *radical* change of consciousness; and it experienced such a transformed consciousness as a gift. What form should and could the radical transformation of consciousness take today and tomorrow? How does a radical change of consciousness relate to the process of healthy maturation?

[1] This text has been translated from Chaps. IV, 7 and 8, in: C.F. von Weizsäcker, *Bewusstseinswandel,* (München: Hanser Verlag, 1988): 244–256. All rights of the Weizsäcker Bequest were transferred to the Carl Friedrich von Weizsäcker Society, a nonprofit society whose executive chairman, Dr. Bruno Redeker, granted permission in March 2014 to use selected texts of Carl Friedrich von Weizsäcker.

K. Raiser (ed.), *Carl Friedrich von Weizsäcker: Major Texts on Religion*, SpringerBriefs on Pioneers in Science and Practice 24,
DOI: 10.1007/978-3-319-03704-2_13, © The Author(s) 2014

With regard to personal consciousness, the process of maturation mostly remains imperfect. Sometimes it fails to the point of catastrophe, of the failure of life. *Healing* would be needed there. Where healing succeeds, it is often experienced as a radical change of consciousness, as the gift of a *conversion.* In the field of politics today, a process of maturation of consciousness is under way. But the analyses of crises (as presented before in this volume on "Transformation of consciousness") leave us worried that this maturation is taking place too slowly, that it will fail in the end.

This worry is not new. In the history of religions, the prophets of the apocalypse expected that the conversion of humankind would only happen as a consequence of a catastrophe of humankind. To someone fully assessing the situation, nothing could seem more appropriate today than such a worry. For years on end, I could not look at a child or a building without the stabbing pain of the question: Will this child still be allowed to grow up? Will this building still be standing the day after tomorrow? And must one not, like the prophets of the apocalypse, place all hope just on a surviving remnant making the conversion of consciousness that, if it were to take place now, would save us all?

Whoever does not know the pain of such questions will perhaps not quite be able to feel what kind of conversion would save us. Whoever knows this pain also knows the inner command not to remain with the pain. For the child who is growing up we must provide daily bread tomorrow and for the day after tomorrow as well. We want to give it an opportunity to grow up not in debilitating horror but in hope that will later give him the energy to collaborate in the conversion. A radical chance of consciousness may today involve just a few people. If implemented reasonably, it will give to many the energy for an evolutive transformation of consciousness. What does it entail?

Religion is unfinished under all four aspects that we have distinguished within it.

Since we have been talking of politics, let us begin with *ethics*. The previous observations (in this volume on "Transformation of consciousness") in response to the question "What is religion?" ended with references to Christianity. When Christians were honest with themselves, they could not be satisfied with the Christian world at any stage of its history. Christianity was an essentially unfinished religion. Only the hope for the return in of Christ could reconcile them with their respective condition life, if at all. The transformation of real history that Christians managed to bring about while living with this tension, which manifested itself from one century to the next, had been overtaken by the Enlightenment at the latest since the 18th century. The Enlightenment has taken over the agency for changing the world. It could step into this role because it took the worldliness of Christian demands seriously. Kant's essay *Towards Perpetual Peace* can be understood as a secularization of the Christian eschatological expectations. The current situation of humankind demonstrates that the demand to overcome the institution of war is a historical necessity.[2]

---

[2] Cf. *Die Zeit drängt*, Munich, 7th ed., 1988, Chap. 10.

Whoever is familiar with these thoughts runs into conflict with a particular tradition within Christianity. Jesus rejected armed struggle for the coming kingdom of heaven, which his 'zealot' Jewish contemporaries deemed necessary. God brings this kingdom. It has already arrived, and it is growing, like a mustard seed growing up into a tree. After Constantine, when the Roman Empire, instead of being destroyed by God, was ruled by Christians who did not know how to bear responsibility for the world otherwise than with weapons, Jesus' words were interpreted as pointing to a future that was dependent on the will of God, beyond the history of this world. 'Chiliasm,' the endeavor to realize in the here-and-now the 1,000-year kingdom of the apocalypse of St. John (*Revelation 20, 2–3*), was rejected as heretical. This conduct by the church leadership was perhaps unavoidable in the unfinished religion of historical Christianity. But I cannot see that in this manner Jesus had been interpreted correctly. He showed to those, who were following him, the way to fight with the weapons of peace for the coming kingdom of peace.

It is correct that this is not a politically planned struggle. Its success lies in God's hands. But that does not mean that in following Jesus' instructions we shouldn't want to be successful in this world. Jesus' message is that we should not trust in our own power as a guarantee of success, that power distorts the will. This is evident in many of his instructions and parables. Only in the post-Constantinian church does a situation arise, where Christianity has no choice but to expect the kingdom of heaven beyond current history, and reward and punishment for one's deeds beyond the grave; this led modern Christianity to expect salvation in the inwardness of the soul. This understanding can also refer to Jesus' words, but not in the same directness with which he spoke to his disciples and the people around him.

I mentioned the Quakers[3] as one example where is has been possible to follow Jesus's words directly also in later Christianity and thereby to achieve limited precise aims, such as, making peace with Indians in Pennsylvania. Mahatma Gandhi is probably the greatest example in our century. People today, who are unable to follow Jesus' words, feel satisfied to point out that the liberation of India was also a limited aim and that the British understanding of the state of law belonged to the conditions for its success; one might also add that it was no longer economically worthwhile for Great Britain to maintain its political rule in India. But this is an objection only to the tendency to confuse the words of Jesus with an infallible means of power. And overcoming the institution of war is also beginning to present itself in our day as a limited precise aim.

The reference to Gandhi points to another problem related to the unfinished character of religion: the diversity of religions. It would be an illusion to think that the religions could be combined 'syncretically' into a single modern "world religion." They share in the multiplicity of cultures. Their diversity is one of the riches of humanity. The question today is only whether, in view of their traditional

[3] Cf. In this present volume "The Sermon on the Mount: Interpretations", p. 72.

antagonisms, the religions will be able at all to contribute toward solving the common problems of humankind.

I shall attempt an answer: The major religions are basically compatible in ethics and inner experience; their compatibility approaches identity at the highest levels, namely, in the purest forms of ethics and in mystical experience. The differences are primarily culturally determined; they range from the order of society and cultic rituals to images expressing ecstatic visions. The irreconcilable antagonisms occur with regard to self-interpretation, i.e. in their theologies, which in each case claim infallible truth.

Gandhi is one example of what has been possible already in our very century. His ultimate motivation is inconceivable without the Indian notion of salvation of the soul (*moksha*). His limited aim, India's freedom and unity, stems from the repertoire of Western political ideals. He himself had linked his world-transforming active charity with what he had learned in England from Christian communities. Aside from demanding of fellow residents of his ashram strict non-violence and a few common religious ceremonies, he only required that they all follow strictly the instructions of their own religions.

Anyone wanting to stimulate the relationships between religions can benefit from considering their differences as a function of their cultural situation. Confucius, Zarathustra, the Jewish pre-exile prophets, Plato, Aristotle and modern Calvinists—they all think in immediately political terms. For, they all have direct access to the centers of political decision-making; their own situations in life are political. Hindu wisdom, the Gnosis of late antiquity, as well as in many instances also philosophy and post-Constantinian Christianity rather think in terms of the after-world or the inwardness of the soul. They all live in societal orders, which they could not change politically. Modern world-wide communication is changing these social backgrounds. Not long ago I witnessed a conversation between an Indian and an American political scientist. The American spoke of the necessity for a deep cultivation of consciousness; the Indian spoke about caring for the poor. The Indian told the American: "You're talking like a Hindu!" Thereupon I said to the Indian: "You're talking like a Christian!" Both accepted this compliment.

If religions want to meet each other, they should not compare their theologies but try instead to do some good together. They would be surprised at what they could learn about themselves in doing so. I dare not predict more.

Nevertheless I shall say something about *theology* in closing these reflections about unfinished religion. I present it as my subjective opinion, without soliciting agreement.

Even the process of the Enlightenment is unfinished. To an unfinished enlightenment an unfinished religion cannot but be unbelievable. Up to this point I have remained largely silent about this critique. However, explicitness is necessary.

The Enlightenment did not bring the three preeminent cultural representations together into unity: morality, theory and art. It seems to me that its effort is unfinished in all three fields. I suggested previously that self-righteous morality becomes evil; self-righteous theory becomes untrue; and self-righteous art loses its substance. There I only used the word 'self-righteous' only for morality. However, it is applicable in all three cases in order to characterize an outstanding element that hides the broad base of reality beneath itself.

Political enlightenment developed a wonderful moral impetus. It exposed the self-righteousness of traditional religion. That is what I primarily mean when I say that it appropriated for itself Christian self-criticism and thereby the energy to transform the world. However, in criticizing something external from which it considers itself to be different, it remains or becomes blind toward its own motivation for power.

Theoretical enlightenment brought humanity the greatest intellectual enrichment of its history thus far. Nothing theoretical today can be taken as serious that avoids being exposed to the criticism of enlightened thinking. Albeit, whatever has not yet conformed to the yes/no decisions of modern science sinks into the obscurity of not-knowledge. How do we learn to perceive what is escaping us that way?

Art offers perception to the world of enlightenment. By stylizing art as one of its values, society retreats, though, from the painful criticism by this perception.

We have described art as a perception of form through the creation of form. In the same manner religion's self-interpretation in myth is such a perception through the creation of form; the same is true for theology—after it has been exposed to the theoretical work of the early Enlightenment. What does this perception teach us?

Myth conveys perception by telling a story. It thus appears as a piece of poetry, as a kind of art. But its 'sacred' character, its religious dignity, distinguishes it from the 'artistic' freedom of modern art. What does it perceive?

A "co-awareness of the whole," a sense of *orientation,* exists in any sophisticated individual perception, even in animal behavior. A cat does not simply see a mouse; it rather sees a mouse scurrying among the blades of grass in the meadow that it is prowling through itself. To a traveler, 'orientating' himself means determining the cardinal points by locating the direction of the sunrise, the Orient (*oriri*: to rise). This co-awareness of the whole mostly remains inexplicit. Reason, defined as perception of the whole, makes this co-awareness explicit.

The grand religious myth is a way to orientate ourselves in life as a whole and in the world. In this sense, myth is the oldest organ of reason. Babylonians gain orientation in the world that Marduk, the god of the empire, has built out of the slain body of the primeval goddess. Jews, and later Christians, orient themselves in the world that the One God had created, like a gardener his own garden.

Myths are as multifarious as the gods they recount about and the situations in which they provide orientation. The great religious thinkers—as in Hinduism, Buddhism and Greek philosophy—thoroughly saw through the metaphorical character of these stories. They spoke in great abstractions about the whole, in which we find ourselves: the One of the Vedanta and Platonism or emptiness of Buddhism. A classical Islamic philosopher, I believe it was Al Kindi (*c. 800–870*), is supposed to have said: "To the question whether the Koran has been revealed by God, the answer is: Yes, it has been revealed by God in order to make the content of philosophy understandable to people who cannot grasp philosophy."

Greek *philosophy* is *enlightenment* by the powerful tool of the *logos*, the proposition deciding about True or False. However, it indicates the perception of the whole as it were in a "myth of the logos"; Plato speaks about the ascent to this perception in the parable of the cave, which allows for a politically ethical, mathematically scientific and meditatively mystical interpretation. Aristotle's god[4] is the eternal mind that knows the everlasting forms of events; he is the unmoving mover of all moving beings; he is the sole full actuality who does not expect any mere potentiality.

Christian *theology* comes decisively *after* Greek philosophy. It is a rationalization of the biblical stories and of Christian experience by means of the philosophical logos. It thus assumes for itself the claim to be true and not false. It sets biblical statements apart as *revelation* (*revelatio*: to reveal; *apokálypsis*: to uncover), and distinguishes them from the myths of other religions and the insights of reason in philosophy. Linguistically therefore, the word 'revelation' denotes itself a perception made possible by the divinely created form of the biblical statement. Dogmatic conciliar decisions passing judgment on heretics become instruments of church politics: unity in the Church through unity in the doctrine.

The intellectual tool of power of *modern enlightenment* is *Science*. Science, as it conceives itself today, is based methodologically on demonstrating the particular, on *experience*. Taken in isolation, this methodological principle cannot uphold the comprehensive claim of unity that we encounter in myth, philosophy and theology. Science does, however, know of a modest co-awareness of a whole by means of *concepts*, without which no experience expressed in scientific terms could exist. It is essential to our present consideration that one can subdivide sciences into two groups, in two somewhat different ways. In terms of *content,* the *natural sciences* can be put over against the *humanities* and the *social sciences.* This division is based on the distinction between nature and humans. It is obviously not precise because a human being is, of course, a child of nature. Medicine is largely the natural science of human beings; animal behavioral science is the social science of animals. *Methodologically*, one distinguishes between *law-based sciences* and the *interpretative sciences. Mathematics* is the most precise instrument of law-based science. The operative tool of interpretative science is the human *understanding* of what is means to be human; this is also called *hermeneutics*, which literally means the science of interpretation. Roughly speaking, natural science is capable and in need of the form of the law. The social sciences, based on understanding, frequently still try to emulate the ideal of a law-based science. The humanities or cultural sciences are inconceivable without understanding (without the ability to communicate—to use a modern term). This methodological distinction is meant when the German generic concept 'Wissenschaft' is not used in current English at all, which rather speaks instead of *science* and the *humanities* (or the *arts*).

[4] Cf. today: G. Picht *Aristoteles »De Anima«*, Stuttgart 1987; E. Rudolph, *Zeit und Gott bei Aristoteles*, Stuttgart 1986.

Historically, both these forms of science critically confronted religion. *Natural science* has drawn a picture of the unity of nature within the scheme of the laws of classical physics, a picture that—as Laplace expressed it to Napoleon—"had no need for the 'god' hypothesis." One then had to invent the Cartesian dualism of matter and consciousness, in order to be able to describe human understanding of what it means to be human; in this scheme, a gap remained for God but no *scientific* necessity to make use of this gap. Great scientists have surely always had a religious reverence for the wondrous order of the laws of nature. They recognized themselves in the contribution by mathematics to the perspectives of classical philosophy; Einstein professed his belief in "the God of Spinoza". However, in comparison with Christianity this is at least a new religion, whether it be 'pantheism' or a reduction of God the Creator to the role of an engineer who had once designed a machine. This controversy stood at the forefront of relations between religion and enlightenment in the centuries of modern times; and it still fills popular polemics today. This is unfinished enlightenment confronting unfinished religion.

A more exact and therefore, at the same time, deeper and more fruitful skepticism with regard to religion was shown by the *humanities and social sciences.* First of all, this kind of science relativizes a traditionalistic justification of specific religions. The dogmas of a religion are not already true simply because of the affirmation that it is the religion of the forefathers. I realized that already as a schoolboy. I grew up as a Lutheran; does this justify me charging my Catholic and Jewish classmates, the Hindus and Buddhists of Asia, the atheists in Marxism and science, with error? This almost self-evident criticism obviously threatens religion's traditional role as bearer of a culture, the validity of which remained essentially unexplained—like the rules of traditional customs. And it also relativizes the specific theological concept of truth of Western tradition (including Islam), which is based on theoretical yes/no decisions. What do we have to say yes and no to? To the absolute truth claim of one religion in order to be able to charge all others with untruth?

This skepticism not only penetrates deeper than the scientific world view; it is also much more productive. The interpretative humanities and the analysis of social situations open up for us the richness and profundity of religious texts and traditions. If I may say so again out of personal experience: I came to appreciate the richness of the Old Testament, when as a young student I read through the Bible and later became acquainted with Old-Testament scholarship and learned to situate the texts within their historical, remote, yet sympathetically human situations. I eventually also succeeded with this in the New Testament as well, with somewhat more trouble, owing to the restraining need by Christian theologians to find absolute truth there. My two chapters on the Bible (i.e. on the biblical story of creation and on the hymn in the letter to the Colossian) reprinted here are an expression of thanks to scholarly biblical criticism for this liberating instruction.

How should we receive this gift? Here it is theologically crucial that the great truths of the religions are not theoretical judgments in the meaning of logic. Myth corresponds to a pragmatic concept of truth: It orientates us so that we can act.

The words of Jesus express the same understanding of truth: "You will know them by their fruits" (*Matt. 7. 16*), and also in the theology of St. John's gospel: "If you continue in my word, … you will know the truth, and the truth will make you free" (*John 8. 31–32*). To "continue in my word" means here to "do what I have told you." Knowing by doing: in the context of ethics this means perception of form through the creation of form. We should recognize religions by their fruits. And we should not judge these fruits on the basis of norms we adopt without explanation out of our own religious or enlightened tradition. Here, too, understanding is called for. How else should we otherwise live together?

Let us return once again to science. A great step that is not yet adequately appropriated philosophically has been taken by *physics of our century*. I give here my personal impression of this field of research which is actually my area of competence.[5] Within the framework of *quantum theory*, the Cartesian dualism of consciousness and matter becomes superfluous. Quantum theory could be compatible with a "spiritualistic monism" that acknowledges a single reality, and calls this reality 'Geist' — mind — in line with classical European philosophy. Hegel expresses this thought abstractly in the statement: "Substance is essentially subject", and Schelling in the statement: "Nature is the mind that does not know itself as mind." Quantum theory would not resist a Buddhist interpretation either. According to quantum theory, a decomposition of the world into separate objects is merely an approximation. This approximation is the precondition for conceptual thought but strictly speaking it is false. Reality is not a prefabricated machine.

It would be premature to equate these possibilities of thought immediately with imagery from conveyed theology. A long path lies ahead of us. But it is an open path.[6]

---

[5] Cf. *The Structure of Physics*, 1985, Chaps. 9–12.

[6] I hope to delve more deeply into this in a future book; now published as: *Zeit und Wissen* (Hanser: München 1992, esp. Chaps. I, 10, pp. 447–522).

Chapter 14
Anxiety

Today I want to address four questions[1]:

1. Do we actually live in anxiety?
2. Are our anxieties justified?
3. What does anxiety have to do with the human condition?
4. How does the church deal with anxiety?

First question: Do we really live in anxiety? The answer is 'Yes'; but is that not rather strange? Should we not be ashamed of our anxiety? Looked at from outside, it seems indeed strange that in this country thousands of people would gather for a church congress under the theme: "In anxieties, and see—we are alive". As if our life had been wrenched from anxiety! What in fact is the real situation of this country? Prosperity is greater now than ever before. What today is called material hardship would not have been described this way before. Physical life expectancy is longer than in all earlier generations. Our constitution respects the rule of law, we have the freedom of speech, we can travel without weapons, and we can cross most borders without a visa. There has been peace for the last 30 years, and we make our plans as if peace will last forever. This is our material reality—and yet: we live in anxiety. Are these fears not an ingratitude toward those who have achieved what we enjoy, an ingratitude toward destiny? Should we not admonish each other to get rid of this unnecessary anxiety?

Yes, we should do that. That is a first partial answer.

But while we are admonishing each other, we must not close our eyes to the fact, that at the same time there are billions of people who do have cause for concrete fear. In Israel and the neighboring Arab countries there is anxiety about

[1] This text is an address at the meeting of the "Deutscher Evangelischer Kirchentag" at Frankfurt/M 1975. It was first published in English in: C.F. von Weizsäcker: *The Ambivalence of Progress*. Paragon House, New York 1988, part I, Sect. 6; the German original is part of C.F. von Weizsäcker, *Der Garten des Menschlichen*. (München: Hanser Verlag, 1977): 122–133. All rights of the Weizsäcker Bequest were transferred to the Carl Friedrich von Weizsäcker Society, a nonprofit society whose executive chairman, Dr. Bruno Redeker, granted permission in March 2014 to use selected texts of Carl Friedrich von Weizsäcker.

K. Raiser (ed.), *Carl Friedrich von Weizsäcker: Major Texts on Religion*, SpringerBriefs on Pioneers in Science and Practice 24,
DOI: 10.1007/978-3-319-03704-2_14, © The Author(s) 2014

the threat of a new war in the Middle East; there is anxiety about the possibility of not surviving this or the next year because of hunger as in India, Bangladesh and in parts of Africa; there is anxiety with regard to the victorious party, to which one has not belonged, as in Czechoslovakia since 1968, in Chile since 1973, in Vietnam and Cambodia since 1975. What emotions are awakened in us by television images that show us scenes from these horrors? "Should one help there?" or even more: "What, if that should happen to us?" In any case these images play on the screen of our conscious and unconscious anxieties.

But we ourselves, do we really live in anxiety? The second partial answer is: "Yes, more than we generally know". Whoever wants to know about the anxieties in his own soul, receives, occasionally perhaps, indications through dreams. I have, next to other dreams that are more deeply rooted in my unconscious, three characteristic dreams of anxiety that infrequently but obstinately return year after year to disturb the quiet hours of my early morning sleep: (1) Hitler is again in power, democratically elected, and has begun his war again, which he will lose again. (2) The Communist secret police sends me to Siberia, and I will never come back. (3) On the bright Western horizon a tiny cloud appears, the first atomic mushroom of the beginning Third World War. These are typical traumas of my generation, and one cannot understand the politics of this generation, if one does not know its anxieties. Actual anxieties can be recognized by all of us from political terms of abuse such as 'communist' or 'radical.' They remind us that our prosperity and the state of law, which represent at the same time an order of social rank and property, are not self-evidently guaranteed. Anxiety is expressed in the aggressiveness of political terms of abuse. Those television images of an as yet spatially distant reality awaken in us the dull feeling that it may be our own situation that is unreal. Aggression makes this discomfort somewhat easier to deal with in that it attributes the threat to others, to the 'guilty,' rather than ourselves, and thus relieves our conscience. And then come confirmations of our fear like kidnappings by political and terrorist groups, burglaries, acts of violence. And how greatly they affect us when they are happening nearby!

I ask for your understanding if I do not leave aside one aspect that is not part of our topic today. Perhaps anxiety, and the distorted behavior due to its hidden presence, manifests itself more strongly in our country than in others with a similar level of prosperity. If this were the case, it would have its origin in our conscience, because since more than 30 years we are suppressing a sense of guilt, and suppressed guilt causes anxiety and other related symptoms. Other peoples may have their problems, this is our problem.

I will not now dwell on a description of the anxieties of everyday life in society, the anxiety to fail in performance, in love, or of fear of the void and of meaninglessness. But these anxieties exist. Do they have a point? Where do they come from?

Second question: Are our anxieties justified? And if they are justified, why is this so? What is the cause of this condition of anxiety? I attempt an answer: "Yes, they are justified." But the reason that justifies our fears lies in the fact that all people act out of anxiety. Anxiety itself creates the conditions and the events about which we are rightly anxious. The system of anxiety stabilizes itself.

In order to examine this response, I propose to divide the question into two parts. A convention of language distinguishes between fear and anxiety. Fear is the emotional response to a definite danger. Anxiety then is a general mood related to fear that cannot point a specific reason for apprehension, but which does not disappear despite sympathetic encouragement.

The response should first focus on fear. The system of apprehensions stabilizes itself. Each partner is afraid, with reason, of the behavior which originates in the fear of the other. In giving initial reasons for such a general thesis, it may be permitted to start from one's own experiences. I would like to elaborate the thesis with the help of two examples with which I had to deal professionally, and still do. The first example is the debate about nuclear power, and the second is the problem of the prevention of war.

In the controversial debate about nuclear power, there exists an apparently great fear on the side of the environmentalists and citizens' initiatives. It is the fear about destruction of the environment and the possible extinction of some or of all of life, beginning with climatic effects caused by all large nuclear power plants and continuing with the possibility of radioactive pollution of air and water. Those who are afraid themselves will often have a distorted view of the motives of their opponents. It is important to recognize that among those, who promote nuclear power, there is also deep fear and not only because of the easily denounced profit motive or the fetish of economic growth for growth's sake. Their fear is that our economy will not be able to compete internationally in the future, unless continually growing sources of energy are available. For those who think farther ahead, this concerns also the question of how hunger and poverty in the world could possibly be eliminated without continued growth of energy. An empirical study of citizens' initiatives has arrived at the conclusion that they are more ready to risk conflict, if they do not have to fear economic disadvantages. The promoters of nuclear power will probably in the end be able to silence their opponents, without resorting to direct repression, by arguing that the opponents' own jobs will be threatened by a lack of energy. But also the pronuclear side has a distorted view of the motives of its opponents. The technocrats refer to antinuclear motives as being 'emotional.' Of course, this depreciation of motives of the opponents to nuclear power has been made easier by the lack of specialized knowledge on their part, causing them to defend their deeply felt concerns by relying on inaccurate specific arguments.

At first glance, the contemporary reality supports the fears of both sides. The structure of our economy is such that without continued economic growth that is dependent upon new sources of energy, there exists the possibility of deep economic crises. On the other hand, continued unlimited growth of energy consumption is climatically insupportable. Also, the issue of the security of a further large scale development of nuclear energy has not been discussed conclusively among the experts, not least in view of politically dangerous times. But on a second approximation, anticipating intellectually future possibilities, it might very well be possible to overcome the threats perceived respectively on both sides. There is no metaphysical reason why a rational future economic use of energy should not be possible. In most cases, there are technological solutions for recognized dangers caused by technology. Elsewhere I will speak publicly about the technical-economic aspect of the problem. Here I am concerned with the human aspect. With regard to dangers caused by human action there is, in the end, no technical solution, only a human one. Why is our economic-technical system constructed in a way that it produces the respective double dangers, i.e. crisis and destruction of the environment?

From the perspective of economic history, growth, also dangerous growth, is a consequence of the system of competition that is generally called market economy by its supporters and capitalism by its critics. Also, the reference to technical progress as the motor of economic growth itself points us back to the competitive system. The reason is that this system has been responsible for profitable technological progress, much more so than the inventiveness of individual researchers, or the mercantilist and socialist systems of centralized state planning. The competitive system grows in its entirety, because the final bonus of survival for those participating in competition depends on their own, private growth. Some refer here to the psychology of the individual entrepreneur or manager; but it is inadequate to derive the characteristics of the competitive system from the profit motive of this social class. A much deeper and much more rational motive for the successful manager is the palpable fear of losing out to competitors. The fear is rational for each individual participant in the system of competition, since he knows that those, who are successful among his competitors, are successful, because they act out of the same sense of apprehension. In this sense, the system, from a psychological perspective, is being stabilized by fear of the consequences of another's fear. Here is the origin of the system's growth and survival power, because the rationally controlled application of this fear and its transformation into aggressive action is a powerful engine. But here is also the origin of its dangers, because who now represents the general public interest that is not represented by private interests, such as the protection of the environment?

The second example, i.e. the prevention of war, I will merely mention. We recognize in it the same structure, only here the competition is not that of economic groups, but rather of sovereign states with their military potential. In classical foreign policy it is rational for each great power to want to be more powerful than the other great powers out of the fear of what the competitor might be driven to do by the same justified fear. The result is periodic power testing, and that means war. The deeply rooted anxiety about a Third World War is certainly well founded.

If, though, fear legitimizes fear, and if fear stabilizes a system of fear, is there a way out from this vicious circle? This introduces another theme that I shall only mention today, namely that of the structure of society. The opinion is widely held that a socialist transformation of the existing social system would be able to break the vicious circle of competitive interests with their rational apprehensions. Is this hope justified? So far at least, revolutionary socialism has led to power structures, in which the secret police personifies the anxiety of the leadership. The controversy as to whether or not this is a necessary attribute of determined socialism refers us—if it is conducted seriously at all—back to anthropological arguments, arguments about the nature of humans as social beings. I will therefore immediately turn to the anthropological question.

Third question: What does anxiety have to do with the human condition? I can again subdivide the response into the separate concepts of anxiety and fear. In my response I will try to offer two definitions that have to be explained subsequently. Fear is the emotional perception of a danger. Anxiety is the fear of our own incapacity for peace. I introduce here two new concepts that need further elaboration: perception and peace.

Any contemporary anthropology must know, as background information, about the evolution of organic life and of animal behavior. What do we mean by perception of a danger? Every living being is threatened. This is something different from the transitoriness of inorganic things. A stone lies for a 1,000 years in one spot and disintegrates; other stones take its place. Organic life has its origin—if I may say so—in an invention of nature, the invention of forms that show the capacity of active self-reproduction and self-preservation. Self-preservation can fail. Therefore every living being is threatened. Self-preservation must fail, for evolution proceeds beyond all of its previously achieved phases. In organic life certain goods are always scarce. Successful evolution makes them scarce, and their scarcity pushes evolution along.

I will not mention the intermediate phases of evolution, but speak immediately of the perception of danger of which human beings are capable. This perception is related to the capacity for imagination. The human person can perceive "something as something," danger as danger. One can even perceive danger when it is not present, and this is the basis for the capacity for human fear. Fear itself is not a rational act; it is emotional. But it is sensible. It has, if I may again use another figure of speech, the rationality of the irrational. The structure of the power system is connected with the limited rationality of fear. Imagination permits us to accumulate means for defense. Against the dangers of nature this accumulation is reasonably limited. But my human competitor can accumulate as much power as I. This compels me to further accumulation. I have described this structure of power systems before with the help of examples. Such power is essentially tragic, because each single step in rational increase inevitably increases the danger for everyone.

Fear and its corresponding power systems are, however, not the final word in human community life. They are only a specific danger linked with the conditions of human development. I would now like to relate the mysterious phenomenon of anxiety, of the inexplicable fear that does not know what it is afraid of, to the perception of this danger that is inherent in human life. I have mentioned before the fear of our incapacity for peace. A peace I would call—if I am permitted to use a third mental image—the embodiment of a truth. Truth, in this context, I would call a perception that is somehow the positive counterpart of a fear: the perception of a reality that makes human life, above all community life, possible—in brief: perception of a possibility to live. Embodiment of a truth means the realization of this possibility. Every perception is incomplete; every human truth is only a partial truth. Therefore, I speak of 'a' peace as the embodiment of 'a' truth. Human history, as we know it, is the history of struggle between truths, struggle between successive forms of intermediate peace. The contemporary problem of peace is, in political terms, the problem of world peace. The struggle between competing truths destroys, again and again, older possibilities for peace. Inevitably this process leads to new anxieties again and again.

In line with presently shared levels of insight, I have tried to explain anxiety with reference to the reality of human life in society. But it is also related to the capacity of the individual to live together with himself, i.e. with the peace of the human person that in a certain psychological mirroring is called his/her identity. The human person certainly gets to know him/herself by living together with others, by being supported by their love, learning language through their language, and finding his/her sense of order through their moral customs. In this respect we are fundamentally social beings. But maturity only develops in the individual. Nothing, which another does for you and for me, nothing which you or I do for another, can release us from the task of becoming ourselves. All great advances in society have been initiated by individuals and have been carried through by individuals. Of course by individuals in society, this means by people who had the capacity for peace in community, because they had successfully achieved a sufficient degree of peace within themselves, a perception of themselves. One cannot sacrifice what one does not have, including oneself. In this sense, anxiety is indeed the perception of the danger of not being capable of peace with oneself—thus, the danger of not being able to perceive oneself with regard to that on which one's own existence depends; or, to express the same insight differently: the danger of not being able to embody self-perception, and thus to be obligated to fear or hate oneself.

Whoever has been trained in depth-psychology could expand this line of thought with many further reflections that would perhaps offer one of the most important topics for this kind of church congress. I cannot enter this field of questions now, but turn to the question that this meeting cannot avoid.

Fourth question: How does the church deal with anxiety? Response: Precisely the peace, about which anxiety is concerned, is the focus of the gospel.

With all that I have said so far, I have basically said nothing. I have explained what we are afraid of, and that we have reason to be afraid. We are afraid of one another and of our own incapacity for peace. Peace is the requirement. But how will we be capable of realizing it? Morality requires it. But does not morality, if we take it seriously, lead us anywhere else but into desperation? The social revolutionaries are morally motivated. Is it mere chance that their creation of new power systems comes about only after much blood has flown? There are instances where, in the historical struggle of truths, we must concede that they are right; that their new power system is different in the one, decisive nuance from the one it has replaced. But it is their tragic, that they ultimately leave unchanged what they were outraged about originally: a system that is being maintained through fear.

I would like to quote a few passages from the New Testament to remind us that it speaks to these questions. "I have told to you all this, so that in me you may have peace. In the world you will have trouble.[2] But take courage; I have conquered the world." (*John 16:33*) Here anxiety and peace are dealt with in the same verse. As a more precise translation the word *thlipsis* that Luther rendered as 'anxiety' means constriction or tribulation, similar to the word *stenochoria* used by St. Paul. In Luther's language, anxiety meant precisely a constriction. It is the constriction that belongs to the condition of the world, the cosmos in the sense of the Gospel of St. John, which has not accepted the light by which it was created. A second passage concerning the conquest of the world says: "Whatever is born of God conquers the world: and this is the victory that conquers the world, our faith." (*1 John 5:4*) A third passage reads, "Now faith is the assurance of things hoped for, the conviction of things not seen." (*Heb. 11:1*) If I read the original Greek with the knowledge of the conceptual language of Greek philosophy and jurisprudence which was familiar to intellectual at the time, then I must translate this passage as follows: "Now faith is the substance of things hoped for, the critical assessment of things that one does not see." The Greek text here uses the term *hypostasis*, the substance, the essence, the thing itself. *Elenchus*, the other decisive term, means the assessment in a court of law or in debate, testing the capacity for resistance.

I will attempt a brief elaboration of these biblical quotations. The world forces us to be anxious. Anxiety is not a weakness of judgment, but rather a correct perception. "I have triumphed over the world," says in the Gospel of St. John the incarnated Logos, the Word of God that has become flesh as it is expressed there; the Logos—through which everything has been created that exists, the essence of the world as creation. Incarnated: that does not mean an abstract metaphysical concept of theology, but means a human being among us, real life in the world. This person is the first born among the many brothers (and sisters): this life is the possibility of life also for us. Seizing this possibility is faith. Faith is initially not a form of conduct, and certainly not something unreasonable: It is content; it is what we believe in. It is the substance of what we hope for, the reality of the possibility. One can only test reality by engaging oneself with it. There is no critical test of a possibility of life without 'testing' it, including the possibility of failure. Faith is the test of those things that one does not see. It brings the truth to the fore by practically engaging itself with peace. That means to embody the soul.

These texts, and the interpretations that they are suggesting to us, may perhaps appear to many of us, as if they were written in a symbolic language to which we are missing the key. One can also illustrate the same experience that is expressed through the interpretation of the Sermon on the Mount on three levels. The first level is the universal ethic of the golden rule, "Do to others as you would have them do to you" (*Matt. 7, 12*). That is the condition under which alone a just society is

[2] The German original here speaks of 'Angst', i.e. 'anxiety'.

possible. It is the condition for external peace in which we can live free from the fear of each other. But who really wants this? Don't we think, "I would want it, but the other does not?" That is the psychological mechanism of projection: I need another subject for my own evil thoughts. Therefore, we come to the second level which stresses that the command of morality concerns the entire person, not just its visible actions. Not only the one who murdered a fellow human person is guilty before the court, but even the one who so feels about another person that in an angry mood he would not regret his death. This is the condition of inner peace, of the maturation of the personality, which alone can assure inner peace. But who is really capable of that? The recognition of that which ought to be, leads us even more into constriction, into the anxiety of desperation about ourselves, because we cannot achieve for ourselves, even with good will, what alone would help us. The process of maturation passes through an experience of death. "Very truly, I tell you, unless a grain of wheat falls into the earth and dies, it remains just a single grain; but if it dies, it bears much fruit" (*John 12:24*). The third level are the Beatitudes, placed at the beginning of the Sermon on the Mount. Blessed are the peacemakers. Blessed are those begging for the spirit. They are blessed here and now. The constriction has opened itself up. There is the experience of the reality of peace that goes beyond the requirements of morality. It does not dismiss morality but fulfills it, in a way different from what any moralist had thought.

What is the use of this for us? It does not make much sense to talk about this, if our transformed self does not radiate the peace. There is no prerogative here for the Christian church. If you show the Sermon on the Mount or the passages from the Gospel of St. John to a knowledgeable Hindu or Buddhist, he will say: "This is true. This has been said by an enlightened one". But whoever understands that this maturation in the human person is the meaning of anxiety, its truth, should not be surprised that anxiety cannot be overcome without this maturation, whatever else may have been improved by progress.

Now should follow the concrete part, addressing the question. How do we deal practically with our anxiety and with that of our fellow human beings? But it is good to end here. The leap into praxis comes early enough for us in our Western culture. Sometimes reflection is more necessary.

Chapter 15
Death

De morte nil nisi bene

Can one dare to talk about death?[1] Will, what I say, withstand my own death? Does it bear up in the light of the death of my neighbor? Does it bear up in the light of the death that surrounds us all and for which we often have more responsibility than we are willing to admit? But, for this very reason: can one dare not talk about death? Can we seriously speak about life, if we do not look death into the eye?

Buddha was the son of a king. It is said that his father raised him in the palace and in the gardens of the palace, far away from the sight of suffering. The prince already had a wife and two small children when his request was granted for the first time to visit the city. We know how starkly human misery is present in such oriental cities. He saw a sick man. "What is that?" he asked his companion in the carriage. "A sick man." "Must I become like that?" 'Yes'. He saw an old man. "What is that?" "An old man". "Must I also become like that?" 'Yes'. He saw a funereal procession. "What is that?" "A dead man". "Must I also become like that?" 'Yes.' He saw a wandering monk whose face was radiating the peace of knowledge. "What is that?" "A monk." "Can I also become like that?" 'Yes.' The prince returned to the palace. That same night he left behind all of his possessions, his father, his young wife, and his children, and wandered into homelessness. After long years of serious wrestling, he found the way to complete enlightenment.

---

[1] This text is based on a lecture at the "Salzburger Hochschulwochen", August 1975. It was first published in English in: C.F. von Weizsäcker: *The Ambivalence of Progress* (New York Paragon House, 1988), part I, section 8; the German original is part of C.F. von Weizsäcker, *Der Garten des Menschlichen* (München: Hanser Verlag, 1977): 145–166. All rights of the Weizsäcker Bequest were transferred to the Carl Friedrich von Weizsäcker Society, a nonprofit society whose executive chairman, Dr. Bruno Redeker, granted permission in March 2014 to use selected texts of Carl Friedrich von Weizsäcker.

K. Raiser (ed.), *Carl Friedrich von Weizsäcker: Major Texts on Religion*,
SpringerBriefs on Pioneers in Science and Practice 24,
DOI: 10.1007/978-3-319-03704-2_15,

Then he taught for 50 years the knowledge about suffering, thirst and awakening for the salvation of all people.

We do not need to go to the Orient to learn this way of thinking about death. From the farewell conversations of Socrates we learn that to philosophize is nothing else but learning how to die. In the parables of Jesus we hear about the rich man, who wanted to build a new barn for his harvest, so that he could say to his soul: "You have ample goods laid up for many years; relax, eat, drink, be merry." But God said unto him, "You fool! This very night your life is being demanded of you. And the things you have prepared, whose will they be?" (*Luke 12, 19f*).

Probably never before has humankind been so perplexed about death. Of course, the answer of the great religions to the question of death was alien to the daily consciousness of humanity at all times. It was presented in mythological colors to make it bearable. In our time even the language of myths is alien to us. Modern consciousness does not understand either the colored or the uncolored answer. But with our questions we must start from current consciousness, if we do not only want to express something friendly (bonum), or even truthful about death, but want to speak 'bene', that is, in a positive manner, and at least sincerely. The hard core of modern consciousness is science. In three respects I will try to talk as a contemporary scholar. First, as a trained scientist: this science also has something important to say about death. Secondly, as someone conversant with the history of ideas, and thus open for the rich experience of all religions, not just our own Christian tradition. And thirdly, in a philosophical way, that is to say in a manner that does not merely list experiences one after the other, or sacrifices one for the sake of another, but rather seeks to think their inner connection.

I want structure this lecture under three questions: Where does death come from? What is the meaning of death in our human life? Where does death lead us?

Where Does Death Come From?

Contemporary science knows a little more about the answer to this question than was known in earlier times. To put it sharply as a thesis: The ego and death are two mutually dependent and meaningful products of organic life. Allow me to explain this sentence slowly, even somewhat pedantically. I combine this explanation with a discussion of previous philosophical-theological ideas.

The framework of historical facts that have been outlined by science is now generally well known. We humans have descended from the animals, and the animals from protozoa. All of us, humans, animals, plants, single-cell organisms, consist physically of molecules. The first forms of life probably evolved from molecules, following the formation of earth, together with its central star, the sun, from a cloud of gas approximately four and a half billion years ago. The sun is one of approximately hundred billion stars of our galaxy, the Milky Way, which is one of a hundred million galaxies that wander through the space, which can still be reached by our instruments. The known cosmos can only have existed for scarcely more than 10 billion years. The beginning is veiled in hypotheses.

Philosophically, the structure of the history of nature is even more important than the framework. All forms in the universe exist in time. Their common attributes are possibility, actuality, and transitoriness. As humans we speak of certain forms as being possible, using conceptual analysis: A star and a quartz crystal are possible agglomerations of material. This formal possibility, controlled by natural law, must meet with real, historical possibility: now we have a cloud of gas which can be transformed into a star, a molten mass which can grow into a quartz crystal. The historical possibility presupposes, what has already become actuality: The gas cloud and the molten mass must already exist. All that has become actual is transitory, for if it were not so, there would be no possibility for the new: The cloud of gas dissolves itself for the most part into stars, the molten mass hardens or flows into larger magma. Possibility is based on the interplay of actuality and transitoriness. If one ventures to designate the historical possibility generally as life, and the transitoriness of the actual generally as death, then death is a precondition of life. The sentence: "Death is her device to have more life" in Goethe's hymn on nature, can already be employed in this quite formal sense. From the perspective of science, its rather enthusiastic sound is no more and no less justified than the gloominess of the sentence "Everything that lives, lives only to die." However, in a more accurate sense, we do not find death as yet in the transitoriness of the inorganic.

The transitoriness of everything that has developed is familiar to all ancient religions and philosophies. What is new in science is the empirical discovery of evolution, of an ever advancing development of always new and always differentiated forms. There has also been evolution in inorganic nature. Newton still thought that it was possible to comprehend the solar system only as the work of a planning intelligence. Today we are convinced that all forms of inorganic nature emerge from one another in accordance with natural laws and that in terms of a statistical average they proceed to greater and greater differentiation. The belief that the so-called blind rule of probability could only create ever-increasing chaos is false. I can summarize here only the results, but not the arguments, of the conceptual assessment of these questions in the following three theses: (1) In a physical system that is far enough removed from statistical equilibrium, an increase in the number of different forms, and accordingly an increasing differentiation, is to be statistically expected. (2) The established equilibrium does generally not destroy the forms, but rather terminates their proliferation; the so-called thermal death does not leave behind a mush but rather a site of skeletons. (3) Only closed systems reach equilibrium; the universe in its entirety is, we assume, not a closed system.

It is the central feature of the phenomenon of evolution that with progressing evolution the number of possibilities increases. It is true that, with the realization of a possibility, everything else that could have been realized becomes excluded. But that, which has newly become actuality, opens up new possibilities, which did not exist before in reality. Through this increase in the abundance of forms, what has become actuality does not entirely disappear. A record of its having been remains in that which it becomes.

The most meaningful evolutionary invention of nature has been organic life. This sentence has been formulated in the personified language that scientists use in their everyday work, in a seemingly light vein. With conceptual caution, one has to call organic life a possibility that has realized itself in a phase of the history of nature. The metaphor of a "new invention of nature" is to imply that it does not just refer to the possibility of a new special form, but to the possibility of new possibilities, of forms that contain within themselves the principle of the creation of ever new forms. We have no empirical reason today to assume the intervention of any other laws than the normal laws of physics and chemistry. In biology today, it is furthermore the fundamental hypothesis of Darwin's theory of evolution that, if organic life has developed once, its further development will occur with statistical necessity ("with probability approaching certainty") in accordance with recognizable principles, and that evolution, again in accordance with these principles, would be sufficient to lead us to an understanding of the factually developed forms of life, if we would only be able to think through the entire complicated processes.

I must now begin to weave into this presentation the thread of examining the opinions of classical philosophy. I admit that I view the currently widespread emotional distrust of Darwin's theory to be a philosophical misunderstanding for which, however, the undifferentiated popular philosophies of many biologists are partially responsible. Darwinism presents the challenge for a still unwritten philosophy of time. Classical Western philosophy contrasts the transitoriness of formed things with the eternity of the forms themselves. The idea of the idea is, according to Plato, the idea of the Good. From the perspective of the transitory thing, the idea, of which it is a part, is its Good, its highest possibility. Ideas are eternal, and thus the highest possibilities of all transitory things are always the same. In my opinion, Plato, against a background of a mathematical, as well as ethical and mystical experience, dared to override the 'pathos' of the transitory with the 'pathos' of the eternal. Time for him is the eternal, in a numerary way progressing image of the Eternal that abides in the One; the circular motion returning into itself is for him the closest approximation of eternity. It is the beauty of the Platonically understood world that there is nothing new in it, for if that possibility existed—a blasphemous thought—then the world would not have been always as close as possible to the eternal Good. This way of speaking has had its impact in the history of thought.

How are we to think now about the possibility of all possibilities, the Good, if there is evolution? In the philosophy that offers itself to us, the possible appears to be related to the future. It would be a rather abstract notion to attribute timelessness to the single conceptually mirrored idea. The real possibilities emerge only from the actuality of their predecessors. The concept of chance that for many is emotionally tainted only says that we must not attribute necessity to the possibilities of the future. We speak scientifically about chance if, in situations that we conceptually characterize as similar, de facto differences occur in different single events. Our conceptual tools only allow us to formulate a prognosis for the relative frequency of the occurrence of different possibilities, that is the prediction of probabilities. The physical laws are nothing other than the general rules in accordance with which the human subject can make its prognoses possible. I shall return to this point in the third part of my lecture. We must now consider the theory of biological evolution, keeping in mind the philosophical questions.

A theory of evolution must explain three empirically established principles: preservation of the individual, preservation of the species, and evolution of the species. The central principle is that of the preservation of the species. The Latin word species is a translation of the Greek eidos and comes from Aristotelian philosophy. Aristotle was a heterodox interpreter of Plato. The Platonic idea is for him the eidos, the form in the things, whatever the metaphor 'in' might mean here. He finds the eternity of ideas realized in the never interrupted presence of things, in which exists the same eidos. The physical descending of like from like ("a human being begets a human being") is the most beautiful model of this thought. No individual is immortal, but the eidos in the individual, the species, is. This philosophical thought is, at the same time, outstanding empirical biology. Modern biology entered the reliable way of science in the 18th century, when it established the universal validity of the physical descent of all individuals of one species from individuals of the same species, consequently, the principle of Aristotle. The slow variation of species that is the basis of evolution is, in accordance with contemporary biology, only a small correction of the principle of the constancy of species. Any explanation of evolution must begin with the explanation of the constancy of species; hence the central importance of the long-sought discovery of the mechanism of reproduction through the double helix of amino acids. But the philosophical motive of Aristotle was turned upside down through the discovery of evolution: The species, too, are not eternal. Even the beautiful must die, in order to make room for a different beautiful; this is true not just for the individual but also for the form.

One must acknowledge that death has entered only with evolution, and this with progressively increasing severity. According to contemporary conceptual models, organic evolution began with molecules that reproduce themselves. Organic evolution begins with the reproduction of a species. There were originally no provisions for the preservation of the individual. Molecules have only their normally limited chemical stability. But now begins between different species, each of which are reproducing themselves, the competition for the limited available resources, for food. The first step of evolution accordingly produces the new phenomenon of the "struggle for survival". Rapid reproduction is an advantage at first, a more economical utilization of food resources. Those species that are more efficient in such operations will gain over the others. This creates a "selection advantage" for certain kinds of behavior and therefore for further development. Accidental or possible mutations, which are almost always harmful, can once in a while still lead to new efficient species, which are then successful. Thus, the first principle, the self-reproduction of a species, leads to the second principle, the further development. In this development follows a chain of "inventions of nature", which are the origin of the principle of self-preservation, thus of the active behavior of the individual for preservation of its own existence. This includes the superior efficiency of individuals that possess metabolism, whose chemical processes are necessary for growth and mobility. The most advantageous material for higher organic chemistry is a substance that is itself organic. This allows consumption of other organisms, therefore animals that live from plants and predators that live from other animals. Individuals now become threatened by other individuals, and self-protection through durability, poison, weapons, size, and mobility brings advantages. Evolution thus produces the scheme of the individual as the last of its three principles. And only individuals that have adapted to self-preservation can really die. For them death is an occurrence, because they are equipped to protect themselves against it. Thus evolution produces death.

Darwin was influenced in his theory of selectivity through the pessimistic economics of Malthus, who taught that living organisms, such as humans, would continually multiply until they reached the minimum level of subsistence. This thought is a glance into an abyss that was always concealed in the European tradition. In Christian Aristotelianism, God appears as a gardener or head of a household, who provides enough for all of his dependents. From the perspective, though, of the individual, Buddha saw much deeper. Even the early Christian tradition, which saw the world itself and the master of this world as evil, was more perceptive. Those indignant about Darwinism are defending an idealized image of the world. Darwinism, too, contained an optimistic belief in progress, but tended to ignore the fact that the development of species and their ecological equilibrium proceeds over the bodies of individuals. One must belong to the victors to be content with the existing beauty of the world. In my opinion, the realism of Darwin contains within it also a challenge for a deeply penetrating philosophy and theology. The opposition between finality and causality is only an apparent opposition. It is precisely the claim of the theory of selectivity that the finalistic description of the behavior of organisms is justified as a phenomenology, because it explains the origins of purposeful organisms. This reciprocal explanation of the initial and final causes, this foundation of the possibility of becoming, is not un-Aristotelian. But it reaches a new level of insight by crossing the boundary of the constancy of species. The theology, though, of a God who sets purposes appears to me to be an anthropomorphism that is just as difficult to combine with the religious doctrine about God's omnipotence, as it is with the philosophical notion in God's omniscience. Purposes are decidedly signs for human orientation whose conceptual meaning contains within it the possibility that they will not be realized. Purposes are concepts of possibilities, therefore anthropomorphic. Whoever makes God directly responsible for purposes, must also make an Anti-God responsible for the unending profusion of failure and suffering, and is consequently forced into theological dualism.

Death is the result of evolution in a still sharper sense than I have previously expressed. An "invention of nature" is the structure that accelerates evolution. Of two somewhat equally endowed species, the one that develops faster will have the advantage. 'Experimentation' with many mutations belongs to this process. Most probably, sexual reproduction, as opposed to the much simpler principle of cell division, is one of such evolution-accelerating structures. It mixes continually new recessive characteristics and endows each species with a bank of latent hereditary factors that give a chance to adjust to changes in the environment. To maintain this continued reproduction, the most powerful behavioral regulator is certainly required, namely, copulation. Thus, we can understand the world-mastering force of sexual love. But the one who copulates has already taken the first step to make himself superfluous. A certain female spider begins to eat her male partner while still copulating. The deep experience of the connection between love and death is not an esthetic error.

The death of the individual is in a still more immediate sense a facilitation of evolution. The short life span of the individual is a selective advantage for the species because it accelerates experimentation of successive generations. The aging of the individual, therefore, is certainly not by accident a compulsory process, a genetically planned progressive sickness. We conceal from ourselves this fact through naive beliefs that the process of aging is perhaps merely an avoidable result of the wear and tear of our physical components. Why should virtually immortal individuals be physically less possible than virtually immortal species? The speed of aging is actually adapted to the speed of ontogeny, the development of the individual. The mayfly can die after the chrysalis stage and one reproductive day. The human being lives longer because its children need it for a longer time.

What Does Death Mean in Our Human Life?

The human being is the creature that knows it must die. Animals have, perhaps, a perception of the proximity of their own death in fear, or a release from their own instinct of self-preservation. But the human being is the first creature that has the reflective awareness of its own death. Perhaps no other awareness has penetrated deeper into human consciousness and thus opened the way to self-perception, than the knowledge of our own death.

The human situation in confrontation with the knowledge of our own death is paradoxical. The manifest bearer of human knowledge is the ego. The ego is the self-perception of the individual as individual. Without reflection, the ego is the steering center of the self-preservation instincts; with reflection, it is the conscious center of all perceptions and knowledgeable decisions; thus the ego is the human instrument of self-preservation, its first servant. The human being is the animal that can say 'I'. But the individual is the real victim of death. By attaining reflective knowledge, the ego attains the knowledge of its own necessary death.

Let us go one step further into detail. The human being has broken up the animal unity of perception and action through the possibility of imagination based upon language. He can imagine that which does not exist, in fantasy, words, concepts, and in reflexion. He can anticipate the future in the form of possibility. Science, whose language I employ, is the unfolding of this human capacity to relate to the future; concepts are possibilities. The capacity to think possibilities also means the capacity to think of dangers and to prepare defenses against them. It means precise fears and the unlimited accumulation of power. Especially the ego of modern Europeans, and through the triumph of our civilization the ego of modern humanity, experiences the accumulation of power in the form of knowledge, of capital, and of weapons, an accumulation that is more powerful than that possessed by any earlier form of life. It is precisely this foresight that has given us all the power on earth, which recognizes the boundary of one's own death. Whose will be the things that you have accumulated?

The most common human form of relating to one's own death is repression. If this were not so, why would we need to be reminded by the Bible? The psychological mechanism of repression is in itself a necessary requirement for survival. Screening and filtering of information is one of the highest accomplishments of our sensory apparatus. It is a precondition for conceptualization. What normally has to be repressed includes the deluge of distant dangers that do not require immediate action. Repression is a guardian of mental health. But its correct use is a task. The more conscious we become, the more essential it is for us to control our own repressions. Truth becomes, for us, a matter of life and death. In this sense, peace is always the embodiment of truth. Our capacity to live with the repression of death is thus an important criterion of our capacity to live, our capacity for peace.

Let us look next at the moment when I do not just know that one has to die, but that I myself must now die. As Jean Paul says, at the moment of birth, an arrow is shot at each person: it flies and flies and reaches him or her at the moment of death. What, if I hear the whizzing of this arrow? Our medicine has a problem here. It has been part of the religious ordering of life not to leave a human person in ignorance in his/her last hour. The person received the consolation of knowing what the doctor, the priest, the relatives knew: that death was nigh. In contemporary medicine, it has become an accepted conviction that it is a requirement of human concern, not to impose the knowledge of the imminent end on a dying person. There is even the practice now of convincing the dying of the contrary.

I have made an experience in this connection. My most dynamic student of philosophy, even physically a symbol of health, had to undergo a medical examination a few weeks after his doctoral exam. Afterwards his wife came to me and asked for my advice. The director of the clinic had told her about the diagnosis of inoperable cancer. He had urged her, not to reveal this to her husband. She said: "I have never lied to my husband. What should I do?" I knew her and answered: "This you can only know yourself. But if you decide to let your husband know, I will assume the responsibility over against the director of the clinic." She told him. When I visited him afterwards, he had been transformed from profound unrest into quiet certainty. He said: "I now know that I have only few months left for straightening out some things in my life, which otherwise I might have postponed for years." Naturally, the things he was referring to concerned fundamentally the relationship to God. For a while, his condition improved even physically. Until his death I have visited him every week, witnessing the oscillation of his conditions. His dying was more real than anything that I had seen.

Medical doctors, whom I have told this story, sometimes speak of exceptional persons, and they can point to many contrary experiences. Certainly, one cannot expect of people who have learned to repress the thought that life means learning how to die, that in the last moment they can make up for a lifetime of neglect. But even these people perceive something that they can understand if one is prepared to help them with it. Often, the physician is the one who is at a loss, or the technical and commercial apparatus of medicine prevents the physician from doing what is possible. Many experiences give testimony of the perception of the nearness of one's own death, even if this cannot be fully understood by the conscious mind. It does not just accompany the limitation of the ego's range of consciousness as the life force recedes. On the contrary, it is the potential of a heightened and liberated sensitivity.

One of the greatest forms resulting from the confrontation with the shadow of death may be the style of the late works of artists. I have the impression that the style of an artist is often more influenced by an awareness of death, than by the mere passage of the years since birth. The quartets of the 55-years-old Beethoven are just as much examples of the style of old age, as are the paintings of the 60-years-old Rembrandt and the 90-years-old Titian, or the development of the character of Prospero in Shakespeare's The Tempest, or the last music of the not-yet-30-years-old Schubert and the 35-years-old Mozart. I do not know if I can define the formal attributes of the mature artistic style: a diminishing sense of the obligation and ambition to be classical, the unbelievable concentration of self-distanced directness—"When I became seventy, I could follow the impulses of my own heart," said Confucius.[2]

I would like to try to give to dying and becoming a meaning that I can unite with evolutionary thought. Sobriety is required. Blissful desire should not be mentioned except by the wise. For almost everyone, the fear of death and the early desire for death are two variations of the same emotional state. They are, as emotional states so often are, omitted acts, perceptions of the incapacity to live. The religious answer to the problem of death, where it is untainted, always contains an inner sobriety. It is clear in its symbols. In the Gospel according to St. John, Jesus speaks of himself when he says, *Very truly, I tell you, unless a grain of wheat falls into the earth and dies, it remains just a single grain; but if it dies, it bears much fruit* (*John 12:24*). He says this to Greeks, and there are commentators who believe that he was referring to the Elysian mysteries. In any case, as in almost all of his parables, he speaks from the peasant experience of life in the country and in nature. The grain of wheat, as an individual, must die in order to produce many grains of wheat. What does that mean in human life?

Development usually does not happen as a continuous increment, but rather in steps. I would like to call this a development from plateau to plateau. Between two plateaus, the development usually goes through a crisis, through the death of a form. In organic life, the species is a plateau. In the environment, to which it has adapted itself, it is capable of harmonic life; it exists in the world in relative peace. Mutations occur frequently, but they are nearly always harmful in this environment and are eliminated because of their failure. However, changes in the inorganic or the organic environment, or improbable new 'inventions', offer other mutants a better chance. In a crisis, which could be fatal, new forms of life appear,

[2] The German original here exemplifies the mature artistic style further by quoting and interpreting two poems by Goethe and Eichendorff; these passages have been left out in the English translation. (K.R.).

as additions to the existing forms or replacing them. This description, like every description, is simplified; there are always transitions, almost unnoticed crises. But the phenomenon of the plateau is quite universal. Also those comprehensive forms of organization that at the outset I called "inventions of nature" can be designated as plateaus, such as: self-reproduction, cells, sexual reproduction, animals, vertebrates, and intelligence.

Also the cultural, social, and political history of humankind shows such plateaus. Today we call the crises between them revolutions: the Neolithic revolution that invented agriculture, the rise of urban civilization, the empires, the great religions, the sciences, the political revolutions of modern times, and the industrial revolution. In the history of science, the concept of scientific revolution has recently been used successfully. Since human cultural history in its single steps is presumably not primarily the consequence, but rather the cause of actual biological changes, the structural relationship between biology and the processes of cultural evolution is even more striking. The reason for this relationship lies, in my opinion, in the structure of time itself. However, I cannot deal with this aspect in my present lecture.

Finally, even the biographical development of the human individual goes through plateaus and crises, and here the crisis is often feared and experienced as death. Here we are dealing with the ego's dying and becoming. Decisive maturity is not reached without experiencing a kind of death. In another lecture about the Beautiful I have listed five objective principles that designate such plateaus: the Useful, the Just, the True, the Beautiful, and the Holy. I have labeled them as manifestations of the Good. They are methods by which humankind perceives that which makes life possible—truths whose embodiment can be a human peace. Every one of these principles designates a quality of experience that was hardly anticipated in the preceding qualities, and which penetrates with violence or gentleness into the seemingly already closed world of experience. In each of these transitions, a re-evaluation takes place of what the ego had previously known, a detachment, a death of gods.

Considering the Useful, the ego triumphs only over the instinct of the moment in favor of planning for future self-preservation. Use and power are accumulated possibilities that serve the ego, they are information. And yet, something of the ego dies in the discovery of the Useful, a childlike state that was more lovable than happy. In the ethical and the Just, the ego finds the possibility and the demand to radically exceed itself. In Truth, humans become aware, above all, of the human will, independent of prosperity and adversity, a releasing sobriety. The Beautiful is a reality beyond self-defense, a blissfulness. In approaching the Holy, the ego hits against a boundary that cannot be crossed. But only one step through the gate, and you find that there never was a boundary.

Now a word about the experience of death in politics. Politics is the planned structuring of human coexistence in society. As such, it has its share of the Necessary and the Useful and the ethically Just. In the tension between these two principles, which to reconcile would be its calling, politics appears to be an as yet unsolved historical task of humanity. Politics has so far been practiced predominantly in the form of—open or concealed—domination. In the struggle for domination and freedom, it develops a kind of Usefulness that we call power. Power is a plateau of human life. As possession, it can be accumulated almost without limits. As a structure, it is self-stabilizing; the only effective means against power is in most cases simply power. As self-protection, it is tragic because of its limitlessness. Who takes up the sword, will die by the sword. If the victor dies in bed, his successors must tremble. War is so far the unsurmounted ultima ratio of power. Wars are fought continually on earth; a Third World War is probable. Politics is accompanied, almost by necessity, by a particular form of lie: Whatever is Useful for one party is presented as the Just. Exactly because of this, the Truth could be actually a healing principle of politics. But who can hear the Truth? Who is mentally capable of wanting to hear it? Anxiety, the unbearable perception of one's own incapacity for peace, is what makes power tragic, for it eliminates seeing and hearing exactly where they would be most necessary. Even the one who screams does not hear, except his own scream. It seems that one must have suffered through the coming war in the depths of one's unconscious, in order to be able, gently and unmistakably, to show those, who unknowingly are preparing it, the small turn of the path that must be taken, in order to avoid the next abyss.

In the human process of maturation, we die many deaths. Always, the realized death is the way to a new life, the gilded death means living under the spell of death. In the end, the flying arrow always reaches the bodily ego—my ego, as we say in equivocal terms.

Where Does Death Lead Us?

With this question, which nevertheless imposes itself on us, we move from the known landscape into the unknown. The consciousness of the ego does not know how to respond to the question of what is beyond its end. Throughout history it has almost always accepted in faith the religious answers to this question. But the ego is not the entire reality of the soul. Religious experience mirrors a reality that transcends the ego. But in the conceptual language of the ego, this experience cannot be expressed as it is. It remains a secret for the ego. In the mythical form, in which it has shaped the universal history of cultures, it is already colored; it is colored by consideration of the receptive capacity of the listeners, and becomes more and more unrecognizable through the process of historical and social filtering; it is colored as well through the historically conditioned capacity for perception of the religious personalities themselves. A consequence of these multiple colorings is the multiplicity of notions about the hereafter that the history of religious teaches us. We should take a brief look at the most important among them.

In most of the primitive religions, there is an otherworldly presence of the deceased, especially of the ancestors. Many primitive religions, and even more so the early high religions, whose political cultures perceive the here and now in terms of monarchical realms, know about a realm of the dead. There emerges the notion that the moral order of the world, whose implementation is the responsibility of the earthly king, will be brought to completion by the otherworldly king. The balance of crime and punishment will only be established in the here-after.

The religion of Israel, as it speaks to us in all but the newest levels of the Old Testament and still determines the characteristic aspects of Judaism, appears like a countermovement to all of these thoughts. Here, the life of humans with God or against God takes place in this world. It is decided in the light of day. To really live means, in the language of the prophets, to live here and now with God. To really die means to be separated here and now from God. The ego cannot escape into the expectation of an otherworldly fulfillment; it has no need for it. Abraham "breathed his last and died in a good old age, an old man and full of years, and was gathered to his people" (*Gen. 25:8*). The ego does not stand alone in this world; the decision for or against God is always made anew by the people collectively. It is the realization of morality as a commandment of the Holy. Thus even modern engaged intellectuals find genuine access to its moral side. But morality without the Holy is not viable; it is a demand without the enabling possibility. The self-consuming effort of plain morality will inevitably become evil or desperate, even if it remains truthful.

The Jewish light of day is a gift to humankind, but it is not the complete reality. Late Judaism incorporated many of the notions of the hereafter from the surrounding peoples. In early Christianity these notions converge with the eschatological expectation of the coming of the messiah into this world. To the coming kingdom of Christ belongs the resurrection of the flesh. This teaching says that what matters is not the abandonment of the body, but rather its recovery. This expectation focusing on the immediate future was not fulfilled, and Christian theology appropriated the entirely different philosophical doctrine of the immortality of the soul. I believe that the Christian historical expectation is the projected sign of an essentially not yet consummated transformation of human history; but that is not our topic today. The question, "Where does the death of the individual take us?" remains a secret also in the Christian tradition.

The most powerful worldwide notion of what lies beyond death is the teaching of the reincarnation of the soul in ever new bodies. Coming out of India, this belief has conquered the greatest part of Asia and appeared in ancient Greece as an Orphic-Pythagorean doctrine. Today, American and European intellectuals, who are religious seekers, are more receptive to this view than to any of the other teachings about the hereafter. According to this doctrine, the balance of crime and its punishment is being established in the eternal here-and-now. Good works lead to good reincarnations, bad lead to bad. Thus, we pay for the acts of our past lives in each current life. In this notion of karma, the moral order of the world operates like a natural law. But if in this way enough has been done to satisfy the requirements of justice, the question of its meaning must comes back with added force. What is the purpose of the eternal thirst, which generates the eternal entanglement in actions and their consequences? When the future Buddha went into homelessness, he was not just concerned about the sufferings of this one life, but with the unending sufferings of the unending lives. He awoke from this cycle. It is not the chain of continual reincarnations but the awakening that is the core of the Buddhist and Hindu experience.

From the critical perspective of the Enlightenment, the very multiplicity of these notions about the hereafter devalues each one of them. They appear as projections of the wishes and anxieties arising in this world. In light of the classical Greek Enlightenment, the classical Greek philosophy and, above all, Plato, asked the question, "What is the true content of the images?" Life, we learn from Socrates in Phaidon, is learning how to die, because death releases us from the chains that bind us to earthly desires. The soul is immortal, because it participates through knowledge in that, which according to its essence, is not subjected to time—the idea. The emotional part of the soul, according to Plato, is mortal. The reasoning part is immortal, because it is part of the eternal reason. But almost everything that Plato says in the dialogues about these things belongs really to the forecourt of philosophy. Did he really believe in the cosmologically depicted myth of the immortality of the individual soul in its journeys through the heavenly cycles and earthly bodies? Or what did this metaphor signify for him? The souls in the hereafter choose themselves the body for their reincarnation, thus their emotional states and their destinies. This is a profound teaching, but who is really the subject of this choice?

The teachings about immortality are being placed into a different light by the theory of evolution, just as is the doctrine about the ideas. Science today is incapable of addressing this subject, not, because it is impossible in principle, but because of its culturally limited perspective. It operates exclusively with an objective approach; it describes objects as they appear to human subjects without reflecting on subjectivity. But I am convinced that science cannot be made consistent, if it excludes what we philosophically call the subject. Hegel's sentence, "substance is essentially subject," exactly expresses this concern, but Hegel's titanic philosophy is not empirical enough. As long as our observations do not include the subjectivity of nature, our science is not completed as science. What would these observations show? The Cartesian duality of two opposing substances, of consciousness and corporeality, cannot be maintained in view of the physics of this century. Matter is, for our physics, that which obeys the so-called physical laws. But these laws define what the subjects can experience. The subjects, however, we ourselves, appear in the world where we make these experiences. As children of evolution, we are part of nature. It seems as if we must say, "The real perceives itself in respectively different shadowings through the eyes of the children of evolution, through the organs that it has produced". Who is the subject? In looking at evolution, the ego itself has appeared to us like a tool. Do we have to say that there is only *one* self? Science leads to the threshold of an experience that is open to meditation but not to reflection. This is reasonable. Conceptual thinking can comprehend that it cannot conceptually designate the origin of its possibility.

Therefore, an academic lecture about death does not conclude with a conceptual statement about where death leads us. Thinking is an activity of the ego that can filter, clarify, and lead to the threshold of experience. Death is a threshold.

Let me conclude with a sentence that I cannot justify through reflection: Work, not bliss, lies beyond death. Bliss exists in that origin of all reality that also has created death.

Carl Friedrich von Weizsäcker Society

WISSEN UND VERANTWORTUNG
Carl Friedrich von Weizsäcker-Gesellschaft e.V.

Knowledge and Responsibility
Carl Friedrich von Weizsäcker Society

Modern science, especially the natural sciences, has given us the power of Greek gods. However, we would need the wisdom of Solomon to use this power sensibly. This is not something we have achieved, but rather it is a task facing us—possibly the single most important task of our time. In 1994, the Carl Friedrich von Weizsäcker Society had 18 founding members. Today, the Society's activities include the organization of international symposia and the development of projects on the decisive challenges of our time.

Knowledge Means Responsibility: Responsibility Needs Knowledge

"*Knowledge and Responsibility*" is our programme in a nutshell. Inspired by the concerns and by the work of Carl Friedrich von Weizsäcker, the Society tries

- to achieve an unbiased and rigorous analysis of our time in five working areas, and
- to develop projects that particularly address the challenges and responsibilities of our time.

"*What must we do*?" is first and foremost a question of insight; but it carries with it the task of furthering insights by gaining them a hearing and weight. Key programme tasks of the *Carl Friedrich von Weizsäcker Society* are therefore, for example, public conferences, expansion of membership, sponsors, partners and friends; but also to strive to develop in the longer term a "network of reason".

K. Raiser (ed.), *Carl Friedrich von Weizsäcker: Major Texts on Religion*, SpringerBriefs on Pioneers in Science and Practice 24,
DOI: 10.1007/978-3-319-03704-2,

Areas of Activities

Physics, philosophy, theology, economics and *altered awareness* are the areas of activity that will be addressed in our projects. Throughout his life as a scholar, Carl Friedrich von Weizsäcker has continued to address these areas. This is one motive for your choice. The second is the way they create our history and our future: nowadays all societies and cultures more or less depend on scientific and technical civilization, up to and including the solution of their economic and social problems. Still, *physics* may be considered as a "key science", *philosophy* as a warning voice, "Do you know what you are saying, and do you know what you are doing?" *Theology* is the effort to understand what religious tradition can teach us for today and tomorrow, *economics* tries to understand social, environmental and political problems. *Altered awareness*, finally, the fifth area of activity, and which pervades all the others, explicitly or implicitly, systematically addresses questions of action and ethical stance in our time.

Address Prof. Dr. Thomas Görnitz (chairman), Dr. Bruno Redeker (executive chairman), Carl Friedrich von Weizsäcker-Gesellschaft e. V., Bielefelder Straße 8, 32130 Enger, Germany.

Website www.CFvW.de and www.CFvW.org/

E-Mail wuv@cfvw.de

Carl Friedrich von Weizsäcker Foundation

Carl Friedrich von
Weizsäcker-Stiftung

CARL FRIEDRICH VON WEIZSÄCKER FOUNDATION

Modern science, especially the natural sciences, has given us the power of Greek gods. However, we would need the wisdom of Solomon to use the power sensibly. This is not something we have achieved, but rather it is a task facing us—possibly the single most important task of our time. The activities of the *Carl Friedrich von Weizsäcker Foundation*, established in 2002, focus on the organization of international symposia, on the preservation and publication of the scientific legacy of Carl Friedrich von Weizsäcker, and on the development of projects on the key challenges of our times.

The Central Guiding Questions

What should we know? What must we do? What may we hope for? Immediately bring to mind Kant's "What can I know? What should I do? What may I hope for?" At the same time they imply a change of perspective towards reason jointly applied to the challenges of our time, the practical problems that humankind faces today:

- Science and technology model a world without borders,
- Innovations, technology and the market drive change in our time,
- The global population is growing and increasingly divided into 'young' and 'old' societies,
- The gap between poverty and wealth widens ever further: locally, regionally and globally,
- The potential for war and terrorism continues to grow, encompassing ethnically and culturally driven conflicts,
- Our use of resources is increasing, placing stress on the biosphere,
- Human power challenges the inherited constitution of nature,
- Overall political order is dominated more and more by the laws of the market,

K. Raiser (ed.), *Carl Friedrich von Weizsäcker: Major Texts on Religion*, SpringerBriefs on Pioneers in Science and Practice 24,
DOI: 10.1007/978-3-319-03704-2,

- Democratic influence on political processes and decisions is waning,
- Ethical stances become relative in the bazaar of opinions.

In the Chap. 8 of his book *Der Mensch in seiner Geschichte* [Humankind in its History] von Weizsäcker reflects on his adaptation of Kant's questions under the heading "Where are we going?": poverty and wealth, war and peace, human beings and nature, the problems are not resolved. But "with jointly applied reason they would be solvable". This is what Carl Friedrich von Weizsäcker has argued for throughout his life as a scholar: not from the perspective of a developed theoretical system but with rationality following the example of everyday speech, "Be reasonable!" And "Our task for today is the global search for truth". And "Reason means recognizing the necessary, and applied in common, to bringing into being what has been recognized as necessary." If we fail to broaden and deepen our understanding of what lies at the core of the challenges of our time as far as we can, there is a constant danger that we might cause more harm than good. "Hope is the perception of the possible" wrote von Weizsäcker in answer to his third question, and at the end of his book he speaks of his hope in these words: "I have tried to speak about what I have experienced. Others may experience other things, more things. They will act."

Address Dr. Bruno Redeker (chairman), Carl Friedrich von Weizsäcker Stiftung, Bielefelder Straße 8, 32130 Enger, Germany.

Website www.CFvW.de and www.CFvW.org/

E-Mail stiftung@cfvw.de

Federation of German Scientists

FGS Federation of German Scientists

The Federation of German Scientists (FGS; German acronym VDW) was founded in 1959 in West-Berlin by renowned nuclear scientists, including Carl Friedrich von Weizsäcker and the Nobel Prize laureates Max Born, Otto Hahn, Werner Heisenberg, and Max von Laue.

Founding members: G. Burkhardt, C.F. v. Weizsäcker. W. Gerlach

Two years earlier this group of experts had become well-known to the public as "Göttinger 18": Nuclear scientists who had publicly argued against a nuclear armament of the German Bundeswehr. Since then the FGS feels bound to the tradition of responsible science. It has nearly 400 members from different fields of the natural sciences, the humanities, and social sciences, so that a large range of topics is approached at a high level of competence. With the results of its interdisciplinary work the Federation of German Scientists not only addresses the general public, but also the decision-makers at all levels of politics and society.

The members of FGS stand in this tradition. They feel committed to taking into consideration the possible military, political, economic and social implications and possibilities of atomic misuse when carrying out their scientific research and teaching.

In Annual Conferences and in interdisciplinary Expert Groups as well as public comments it addresses issues of science and technology on the one hand, and peace and security policy on the other. At the same time, the role of science itself in genesis and in solution of socio-technological problems is subject of examination and expertise. FGS' membership lists also include representatives of the humanities and social sciences, so that a large range of topics is approached at a high level of competence. With the results of its interdisciplinary work the Federation of German Scientists not only addresses the general public, but also the

K. Raiser (ed.), *Carl Friedrich von Weizsäcker: Major Texts on Religion*, SpringerBriefs on Pioneers in Science and Practice 24,
DOI: 10.1007/978-3-319-03704-2, © The Author(s) 2014

decision-makers at all levels of politics and society. According to its statutes of 1959, the FGS aims to

- keep up and deepen the awareness of those working in science for their responsibility for the effects which their work has on society;
- study the problems which result from the continuous development of science and technology;
- assist science and its representatives in making public the questions related to the application of scientific and technical developments;
- provide advice and thus exercise influence on decisions as long as they are assessable and can be dealt with by means of scientific knowledge and methods, and to point out all forms of misuse of scientific and technical results;
- to defend the freedom of scientific research and the free exchange of its results and to expand and strengthen the traditional international cooperation of scientists.

Carl Friedrich von Weizsäcker had been part of the famous "Göttinger 18", the group of renowned nuclear scientists who publicly opposed a possible nuclear armament of West Germany in the 1950s, and was among the founding members of the Federation of German Scientists in 1959. His spirit, his way of perceiving the world and his understanding of the role and responsibility of science for society and the development of humankind profoundly shaped the self-perception and sphere of influence of the FGS in its early years and later on. He also repeatedly served in public functions of the FGS, most notably as its chairman from 1969-1973.

Address Vereinigung Deutscher Wissenschaftler (VDW), Marienstr. 19/20, 10117 Berlin, Germany.

Email info@vdw-ev.de;

Website http://www.vdw-ev.de/index.php/de-DE/

VDW Vereinigung Deutscher Wissenschaftler

Udo Keller Foundation

UDO KELLER STIFTUNG
FORUM HUMANUM

Mission Statement

The name reflects the programme of action. The Foundation, set up by the Hamburg businessman Udo Keller, sees itself as a Forum Humanum—a forum for all those who would like to investigate the question of the truly human. At a time when technology and economic processes are increasingly influencing human choices, the Foundation addresses the importance of the moral and religious heritage of human cultures worldwide. The Foundation assumes that the future development of human beings will decisively depend on whether we succeed in harnessing the rich potential of these traditions for the future. In this way the Udo Keller Foundation argues for a revival of the question of the purpose of human life in twenty-first century terms.

Funding Priorities

The Udo Keller Foundation Forum Humanum contributes to an interdisciplinary dialogue between natural sciences and the humanities as well as to the multi-faith dialogue between world religions. These goals are being realized at its headquarters in Neversdorf near Hamburg and at its study centre in Tübingen, the *FORUM SCIENTIARUM at the Eberhard Karls University of Tübingen.*

K. Raiser (ed.), *Carl Friedrich von Weizsäcker: Major Texts on Religion*,
SpringerBriefs on Pioneers in Science and Practice 24,
DOI: 10.1007/978-3-319-03704-2, © The Author(s) 2014

Funding Activity

The Udo Keller Foundation Forum Humanum is a co-founder of the interdisciplinary project *FORUM SCIENTIARUM at the Eberhard Karls University of Tübingen* and is one of several inaugurators of the *Academy of World Religions* at the University of Hamburg. The Foundation has sponsored the *Verlag der Weltreligionen* (World Religions Press) since its establishment in 2007, and has initiated various lecture series in Hamburg and Tübingen—including *Thinking the future* (ZUKUNFT denken) in Hamburg in cooperation with the Hamburg Planetarium (2010–2014) and the *Unseld Lectures* at Tübingen (from 2008). Together with the German Literary Archives in Marbach, the Foundation has funded since 2008 the *Udo Keller Scholarship for Contemporary Research into Religion and the Modern Age*.

Additional information on the work of the *Udo Keller Foundation Forum Humanum* may be accessed in German on its website at: www.forum-humanum.org.

Address Udo Keller Stiftung Forum Humanum, Kleine Seestr. 24, 23816 Neversdorf, Germany.

Email info@forum-humanum.org

Ruhr University Bochum

RUHR
UNIVERSITÄT
BOCHUM
RUB

Portrait

Located in the midst of the dynamic, hospitable metropolitan area of the Ruhr, in the heart of Europe, the Ruhr-Universität Bochum (RUB) with its 20 faculties, RUB's disciplinary institutional units, is home to 5,600 employees and over 41,000 students from 130 countries. All the great scientific disciplines are united on one compact campus.

The RUB is on its way to becoming one of the leading European universities of the 21st Century. Almost all courses are offered as Bachelor and Master degree programmes. Our excellence programmes have made themselves an international name: Our Research School is an international college for structured doctoral research in the life sciences, natural sciences, engineering, the humanities and social sciences. Interfaculty and interdisciplinary Research Departments, which are mutually, nationally and internationally networked, sharpen the profile of the RUB. Added to this is an unsurpassed programme for the promotion of Early Career Researchers, and an excellent infrastructure.

What makes it all come alive is the people who meet on campus with their thirst for knowledge, their curiosity, and their commitment. They help shape the RUB and their open-mindedness makes the RUB an attractive place for people from around the world.

Guiding Principle of the RUB

The trio of values, people-centred—cosmopolitan—high-performance, represent the cornerstones of the RUB environment. This space is more than just the sum of its individual elements: Pepople-centred and cosmopolitan means to respect diverse cultures and to give guests a home. People-centred and high-powered means jointly developing creative forces, to 'tackle' things with verve and ambition.

"Campus RUB" is the contemporary universitas—the community in which people take centre stage.

K. Raiser (ed.), *Carl Friedrich von Weizsäcker: Major Texts on Religion*,
SpringerBriefs on Pioneers in Science and Practice 24,
DOI: 10.1007/978-3-319-03704-2, © The Author(s) 2014

Living Universitas

The members of the universitas teach others and, at the same time, learn from each other—whether in science, studies, engineering, or management. Universitas promotes and demands codetermination from everyone—a far cry from academic or hierarchical structures. Living universitas is: When students in the eTutoring project plan courses; when the promotional funding line of the research fund is judged solely by doctoral students; the uncomplicated interaction of researchers from various disciplines.

Address Ruhr-Universität Bochum, Universitätsstraße 150, 44781 Bochum, Germany.

Website http://www.ruhr-uni-bochum.de/universitaet/index_en.html

Email international@rub.de

Prof. Dr. Carl Friedrich von Weizsäcker with his wife in the Alps (1989). © The Weizsäcker Family represented by Dr. Elisabeth Raiser who granted permission to use this photo.

About the Author

Carl Friedrich Freiherr von Weizsäcker (June 28, 1912–April 28, 2007) was a German physicist and philosopher. A member of the prominent Weizsäcker family, he was son of the diplomat Ernst von Weizsäcker, elder brother of the former German President Richard von Weizsäcker, father of the physicist and environmental researcher Ernst Ulrich von Weizsäcker, and father-in-law of the former General Secretary of the World Council of Churches Konrad Raiser.

Born in Kiel, Schleswig-Holstein, he was raised in Stuttgart, Basel, and Copenhagen. From 1929 to 1933, Weizsäcker studied physics, mathematics and astronomy in Berlin, Göttingen and Leipzig supervised by and in cooperation with Werner Heisenberg and Niels Bohr, among others. The supervisor of his doctoral thesis was Friedrich Hund.

Weizsäcker made important discoveries in theoretical physics regarding the masses of atomic nuclei, energy production in stars from nuclear fusion processes, and on planetary formation in the early Solar System. During World War II he participated in the German program for developing nuclear energy and atomic bombs.In his later career, he focused on philosophical and ethical issues, and was awarded several international honors for his work in these areas.

Work on nuclear physics Weizsäcker's special interest as a young researcher was the physics of the atomic nucleus. Simultaneously with Hans Bethe he found a mechanism or pathway for the cyclic process of fusion in stars (Bethe-Weizsäcker process, published 1937–1939). This discovery should not be confused with his 1935 development of the Bethe-Weizsäcker formula, or Semi-Empirical Mass Formula (SEMF) for nuclear masses, again simultaneously with Hans Bethe.

K. Raiser (ed.), *Carl Friedrich von Weizsäcker: Major Texts on Religion*,
SpringerBriefs on Pioneers in Science and Practice 24,
DOI: 10.1007/978-3-319-03704-2,

Work on planetary formation In 1938, Weizsäcker developed a theory of the formation of the Solar System, based mainly on considerations of turbulent motion of gases and dust. The theory also helped to explain the empirically observed regular pattern of increase in the diameters of the orbits of the planets of the Solar System, from inward to outward.

Work on atomic weapons As a theoretical physicist, Weizsäcker (and by his own estimate, 200 other physicists) had recognized immediately after nuclear fission had become known (by Otto Hahn) in 1938 that nuclear weapons could potentially be built. He discussed the upsetting implications in February 1938 with philosopher friend Georg Picht.

During World War II, Weizsäcker joined the German nuclear energy project, participating in efforts to construct an atomic bomb. For some time he had been hoping for political influence growing out of participation in a successful nuclear weapons project. In July 1940 he was co-author of a report to the Army on the possibility of "energy production" from refined uranium. The report also predicted the possibility of using plutonium for the same purpose including the production of a new type of explosives. During summer 1942 Weizsäcker drafted a patent on a transportable "process to generate energy and neutrons by an explosion…e.g., a bomb", which was never filed. The draft was found in the 1990s in Moscow.

Historians have been divided as to whether Heisenberg and his team were sincerely trying to construct a nuclear weapon. In a 1957 interview with the German weekly Der Spiegel, Weizsäcker frankly admitted to the scientific ambitions of those years: "We wanted to know if chain reactions were possible. No matter what we would end up doing with our knowledge—we wanted to know." Weizsäcker said that they were spared the decision on building the bomb as they saw rather soon that the German war economy was unable to mobilize the necessary resources.

Weizsäcker worked later during the war as a professor in Strasbourg. The American capture of his laboratory and papers there in December 1944 revealed to the Western Allies that the Germans had not come close to developing a nuclear weapon.

Postwar career In 1946, Weizsäcker became director of the department for theoretical physics in the Max Planck Institute for Physics in Göttingen. Weizsäcker felt that the scientists who had developed the foundations of such powerful theories as that of the atomic nucleus, should take on the responsibility for the consequences. In 1957, it was mainly he who formulated the protest of the "Göttinger 18", a group of prominent German physicists, against the idea that the West German armed forces should be equipped with tactical nuclear weapons. He suggested that West Germany should declare its definitive abdication of all kinds of nuclear weapons. From 1957 to 1969, Weizsäcker was professor of philosophy at the University of Hamburg. From 1970 to 1980, he was head of the Max Planck Institute for the Research on Living Conditions in the Modern World in Starnberg. He researched and published mainly on philosophy and foundations of physics, but also on the danger of nuclear war, which he thought underestimated by the public and the political establishment, on the conflict between the First World and the Third World, and the consequences of environmental degradation, and on the world as an interlocking whole ('Weltinnenpolitik'). In the 1970s he founded, together with the Indian philosopher Pandit Gopi Krishna, a research foundation "for western sciences and eastern wisdom",

After his retirement in 1980 he intensified his work on the conceptual foundations of physics and on philosophical issues. In the 1980s he invested much of his creative energy in the promotion of what was originally called a "Council for Peace". The movement resulted in the "World Convocation on Justice, Peace and the Integrity of Creation" in Seoul in 1990.

Weizsäcker developed the theory of ur-alternatives (archetypal objects), publicized first in his book Die Einheit der Natur (1971; English translation "The Unity of Nature", 1980) and further developed through the 1990s. The theory axiomatically constructs quantum physics and uses it to discuss the foundation of a universal physics on the quantum mechanics of binary alternatives. Weizsäcker used his theory, a form of digital physics, to derive the 3-dimensionality of space. The program has not, so far, come to an end. In 2007, Weizsäcker died at the age of 94 in Starnberg, Germany.

Awards and honors Max Planck Medal (1957), Goethe Prize of the city of Frankfurt am Main (1958), Pour le Mérite for Science and Art (1961), Peace Prize of the German Book Trade (1963), Erasmus Prize of the city of Herdam (1969), Austrian Medal for Science and Art (1969), Grand Merit Cross with Star and Sash of the Federal Republic of Germany (1973) Ernst Hellmut Vits Prize of the University of Münster (1982), Heinrich Heine Prize of the city of Düsseldorf (1983), Sigmund Freud Prize for Scientific Prose (1988), Templeton Prize for "Progress in Religion" (1989), Theodor Heuss Prize "for his world-renowned, diverse and dedicated contributions to humanity themes: peace–justice–Integrity of Creation" (1989), Prix Arnold Reymond (University of Lausanne), Hanseatic Goethe Prize, Karl IV Prize of the City and University of Prague.

Honorary degrees Law: Free University of Amsterdam, University of Alberta, University of Aberdeen; Theology: University of Tübingen, University of Basel;

Science: Karl Marx University, Leipzig; Philosophy: Berlin Institute of Technology, University of Aachen

Memberships Max Planck Society for the Advancement of Sciences, German Academy of Sciences Leopoldina, Göttingen Academy of Sciences, Saxon Academy of Sciences, Austrian Academy of Sciences, Bavarian Academy of Sciences, Bavarian Academy of Fine Arts, German Physical Society, Académie des Sciences Morales et Politiques, American Physical Society, Croatian Academy of Sciences and Arts, German Academy for Language and Literature, Joachim-Jungius Society of Science /Hamburg Academy of Sciences, Hamburg Institute for Human Sciences.

Among his major publications are: *Zum Weltbild der Physik* (Leipzig 1946, 2002, 14th edition, renewed and with introduction by Holger Lyre) [*The World View of Physics* (London, 1952)]; Le Monde vu par la Physique (Paris 1956); Der begriffliche Aufbau der theoretischen Physik (Lecture Notes 1946) (Stuttgart 2004); *Die Geschichte der Natur* (Göttingen 1948) [*History of Nature* (London 1951)]; *The Relevance of Science (*London–New York 1964); [*Die Tragweite der Wissenschaft* (Stuttgart 1990); *Die Einheit der Natur* (Munich 1971) [*The Unity of Nature* (New York 1980); *The Biological Basis of Religion and Genius*, Gopi Krishna (New York 1971), intro. by Carl Friedrich von Weizsäcker, which is half the book; *Wege in der Gefahr* (Munich 1976); *The Politics of Peril* (New York 1978); *Der Garten des Menschlichen* (Munich 1977) [*The Ambivalence of progress*, essays on historical anthropology (New York 1988)]; *Deutlichkeit: Beiträge zu politischen und religiösen Gegenwartsfragen*, (Munich 1978); *Der bedrohte Friede* (Munich 1981); *Wahrnehmung der Neuzeit* (Munich 1983); *Aufbau der Physik* (Munich 1985) [*The Structure of Physics* (Heidelberg 2006)]; *Die Zeit drängt* (Munich 1986); *Bewusstseinswandel* (Munich 1988); *Der Mensch in seiner Geschichte* (Munich 1991); *Zeit und Wissen* (Munich 1992); *Große Physiker* (Munich 1999). See also the website on this book on Carl Friedrich von Weizsäcker, at: <http://www.afes-press-books.de/html/SpringerBriefs_PSP_C.F.v._Weizsaecker.htm>.

Prof. Dr. Carl Friedrich von Weizsäcker (1989). With his wife receiving the Templeton Prize for "Progress in Religion" (1989) in London. *Source* Templeton Prize — Clifford Shirley; permission to use this photo was granted

About the Editor

Prof. Dr. Konrad Raiser (born 1938 in Magdeburg, Germany) is a former General Secretary of the World Council of Churches (WCC) and Professor (em.) for Systematic Theology and Ecumenics at the Ruhr University Bochum. He studied Protestant theology in Tübingen, Heidelberg and Zürich from 1957–1963 and sociology and social psychology at Harvard University, Cambridge, Massachusetts (USA) in 1965/1966. From 1967–1969, he was assistant in Practical Theology at the Protestant Theological Faculty in Tübingen (Germany). He earned a doctorate in theology in 1970 from the same faculty in Tübingen. In 1969, he joined the staff of the World Council of Churches in Geneva (Switzerland), first in the Commission on Faith and Order and then as Deputy General Secretary and Staff Moderator of the Unit on Justice and Service. From October 1983 to February 1993, he was Professor of Systematic Theology and Ecumenics at the Protestant Theological Faculty of the University of the Ruhr in Bochum (Germany). He assumed leadership of the WCC as General Secretary in January 1993 and retired from this office in December 2003. Since 2004 he lives in retirement in Berlin, Germany. His wife Elisabeth is the daughter of Carl Friedrich von Weizsäcker. They have four sons.

Konrad Raiser is the author of nine books (Identität und Sozialität, 1971, Ecumenism in Transition, 1991, Wir stehen noch am Anfang, 1994, To Be the Church, 1997, For a Culture of Life, 2002, Hoffen auf Gerechtigkeit und Versöhnung, 2002, and Schritte auf dem Weg der Ökumene, 2005, Religion-Macht-Politik 2010 (engl. Religion- Power-Politics, 2013), Ökumene unterwegs zwischen Kirche und Welt, 2013. In addition, he edited five other books. Since 1970 he has published numerous articles and essays on theological and ecumenical subjects, including four entries in the *Dictionary of the Ecumenical Movement* (WCC Publications, 1991), for which he was also a member of the editorial board. He was co-editor of two quarterly journals: *Ökumenische Rundschau,* and *The Ecumenical Review*. He holds honorary doctorates from the Budapest Theological Academy (1992), the University of Geneva (1996), and the University of Hamburg (2009). As guest lecturer he taught at the universities of Jena, Fribourg, Hannover, the Episcopal Divinity School and Harvard Divinity School in Cambridge/Mass., as well as the Lutheran Theological Seminary at Hongkong.

K. Raiser (ed.), *Carl Friedrich von Weizsäcker: Major Texts on Religion*,
SpringerBriefs on Pioneers in Science and Practice 24,
DOI: 10.1007/978-3-319-03704-2, © The Author(s) 2014

Prof. Dr. Carl Friedrich von Weizsäcker (1998). © The Weizsäcker Family represented by Dr. Elisabeth Raiser who granted permission to use this photo.

About the Book

This book presents a collection of texts by the German physicist and philosopher Carl Friedrich von Weizsäcker (1912–2007) in English, for use in seminars on the philosophy of religion, the comparative study of religion, but as well on the relationship between religion and the scientific worldview. Some of the selected texts appear in English for the first time. Weizsäcker became famous through his works in physics, mainly in the early development of nuclear physics. Later he would also become well known as a philosopher and analyst of contemporary culture. He also worked very intensely on projects for the prevention of nuclear war and for peace in general. See also the website on this book on Carl Friedrich von Weizsäcker, at: <http://www.afes-press-books.de/html/SpringerBriefs_PSP_C.F.v._Weizsaecker.htm>.

K. Raiser (ed.), *Carl Friedrich von Weizsäcker: Major Texts on Religion*,
SpringerBriefs on Pioneers in Science and Practice 24,
DOI: 10.1007/978-3-319-03704-2,

The manufacturer's authorised representative in the EU is Springer Nature Customer Service Centre GmbH, Europaplatz 3, 69115 Heidelberg, Germany. If you have any concerns regarding our products, please contact ProductSafety@springernature.com

Printed and bound by CPI Group (UK) Ltd, Croydon, CR0 4YY
22/09/2026
02221059-0019